D0007826

Solid State Devices, 1979

Solid State Devices, 1979

Nine invited papers presented at the Ninth European Solid State Device Research Conference (ESSDERC) and the Fourth Symposium on Solid State Device Technology held at the Technical University of Munich, 10–14 September 1979

Edited by H Weiss

Conference Series Number 53

The Institute of Physics
Bristol and London

CODEN IPHSAC 53 1-153 (1980)

British Library Cataloguing in Publication Data

Solid state devices, 1979. – (Institute of Physics. Conference series; no. 53 ISSN 0305-2346).

1. Semiconductors – Congresses

I. Weiss, Harold

II. European Solid State Device Research Conference, *9th, Technical University of Munich, 1979*

III. Symposium on Solid State Device Technology, *4th, Technical University of Munich, 1979*

621.3815'2 TK7871.85

ISBN 0-85498-144-6
ISSN 0305-2346

The Ninth European Solid State Device Research Conference and the Fourth Symposium on Solid State Device Technology were sponsored by the European Physical Society, The Institute of Physics, Region 8 of the Institute of Electrical and Electronics Engineers, the Deutsche Physikalische Gesellschaft and the Nachrichtentechnische Gesellschaft im VDE.

Organising Committee

H Weiss (*Chairman*), P Balk (*Co-Chairman*), R Müller (*Vice-Chairman*), M Zerbst (*Program-Chairman*), E Lange

Program Committee ESSDERC 79

H Berger, J Borel, P A H Hart, J Hesse, W Heywang, F Leuenberger, R van Overstraeten, F Paschke, G Soncini, P Weissglas, K W Gray, N C de Troye

Corresponding Members

G H Schwuttke, F M Smits, T Sugano

Program Committee of the Symposium

P Balk, K J S Cave, A Goetzberger, V LeGoascoz, J F Verwey, M Zerbst

Honorary Editor

H Weiss

Published by The Institute of Physics, Techno House, Redcliffe Way, Bristol BS1 6NX, and 47 Belgrave Square, London SW1X 8QX, England.

Set in 10/12 pt Press Roman by DJS Spools Ltd, Horsham, Sussex, and printed in Great Britain by J W Arrowsmith Ltd, Bristol.

Preface

There is an increasing degree of interdependence between semiconductor technology and device research; together they stimulate research and development work aimed at improving all aspects of the performance of integrated circuits. In recognition of this the Ninth European Solid State Device Research Conference (ESSDERC 79) and the Fourth Symposium on Solid State Technology were held jointly at the Technical University of Munich, 10–14 September 1979.

Amongst over 120 papers presented during the five days of the meetings were the nine invited reviews which are reproduced in this volume. The interaction of device research and materials technology is clearly shown in the review of trends in MOS circuits where higher degrees of integration are being achieved. Other papers highlight modern fields of application for electronic devices, e.g. in optical communications, gigabit logic and medicine. The growing number of possible applications for semiconductor devices will not be realised without further development of materials like SiC for blue LEDs or cheap silicon for solar cells, or of modern methods in semiconductor technology. In this latter context the three papers on laser annealing, dry etching, and the role of Cl in the thermal oxidation of silicon are particularly significant.

It is hoped that the reader will gain some knowledge about the state of the art in solid state devices.

H Weiss

Contents

Devices for optical communications

H Kogelnik
Bell Telephone Laboratories, Holmdel, New Jersey 07733, USA

Abstract. A review will be given of recent progress in device research for optical fibre communications. Recent results include improved growth of epitaxial layers by liquid-phase, vapour-phase and molecular beam epitaxy in both the GaAlAs and the InGaAsP materials systems, improved structures, performance and life of laser and detector devices, laser structures for single-mode operation, and devices operating at the longer IR wavelengths where fibre losses and fibre dispersion are low. Experimental modulators, switches and filters in the guided-wave form of integrated optics will also be discussed.

1. Introduction

In recent years we have witnessed considerable research effort and rapid progress in the new technology of lightwave communications (for a recent review, see Li 1978). This technology is based on generation of signals in tiny semiconductor lasers or LEDs and on the transmission of optical signals via hair-thin glass fibres. It has been stimulated by many promises, including promises of a transmission medium offering larger bandwidth, larger repeater spacings, smaller size, lower weight, lower crosstalk and potentially lower cost than the traditional media such as copper wire pairs, coaxial cables or hollow metallic waveguides. Telecommunications engineers are exploring a large variety of applications for this new technology. These range from short data links interconnecting computers or other equipment, via systems for television, cable television, and telephone loops, for electric power companies and the military, all the way to telecommunication trunks between central offices in a city, between cities, and undersea cables linking continents. Characteristic of the progress in optical fibre technology is the enormous reduction of the transmission loss of fibres accomplished in the last decade. This is illustrated in figure 1.

A first generation of lightwave systems has already entered the stage of field tests and trial systems carrying commercial traffic. There are now over one hundred installations in Europe, Japan and North America, and among them are some major field experiments. Examples for the latter are the recent Atlanta Fiber System Experiment (Jacobs and Miller 1977, *Bell Syst. Tech. J.* 1978) and the subsequent Chicago Lightwave Communications Project put in operation by the Bell System in May 1977 and reporting extremely encouraging results (Schwartz *et al* 1978).

The first generation of lightwave systems uses multimode fibres of fused silica (SiO_2) doped with GeO_2, B_2O_3 or P_2O_5. Transmission is mostly in digital form with data rates between one and several hundred Mbit/s, and with repeater spacings of several kilometres.

0305-2346/80/0053-0001 $03.00

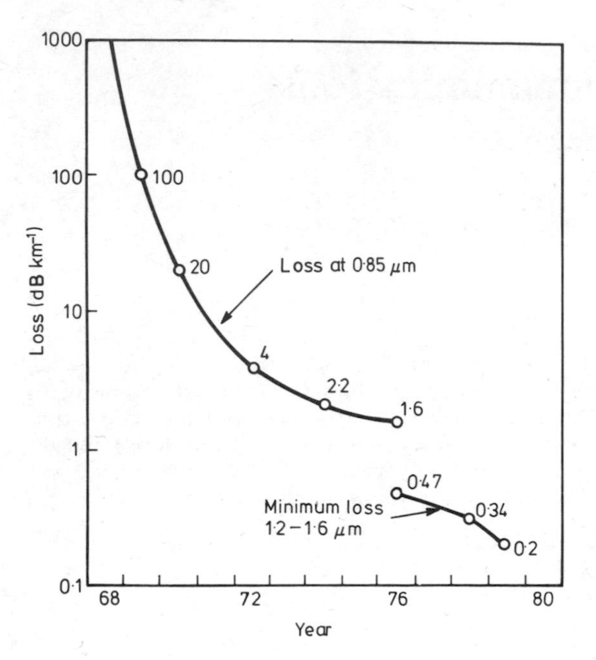

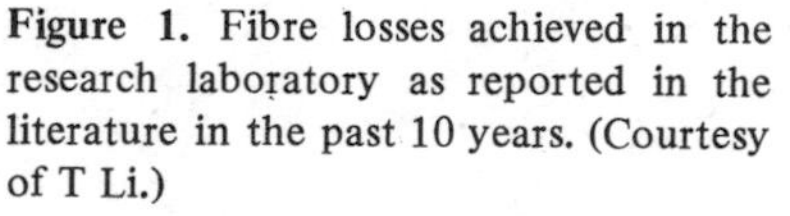
Figure 1. Fibre losses achieved in the research laboratory as reported in the literature in the past 10 years. (Courtesy of T Li.)

The wavelength of the transmitted light is near 0·8 μm. The devices used in these systems are AlGaAs junction lasers or LED sources, and Si p–i–n or avalanche photodetectors.

The transmission speeds and repeater spacings of fibre systems are limited by fibre loss and by pulse broadening due to pulse dispersion in the fibre. Apart from the transmission window near 0·8 μm, fused-silica based fibres exhibit two other important windows near 1·3 and 1·6 μm. As Rayleigh scattering decreases strongly with reciprocal wavelength, fibre losses can be considerably lower in these 'long-wavelength' windows. Recent reports indicate losses as low as 0·6 dB km^{-1} near 1·3 μm and 0·2 dB km^{-1} near 1·6 μm (Miya *et al* 1979). The loss spectrum of such a fibre is shown in figure 2. The dispersive properties of the fibre material contribute importantly to pulse dispersion in optical fibres. This material dispersion is much smaller in the long-wavelength region, where 'zero-dispersion' behaviour has been predicted for wavelengths near 1·3 μm (Li

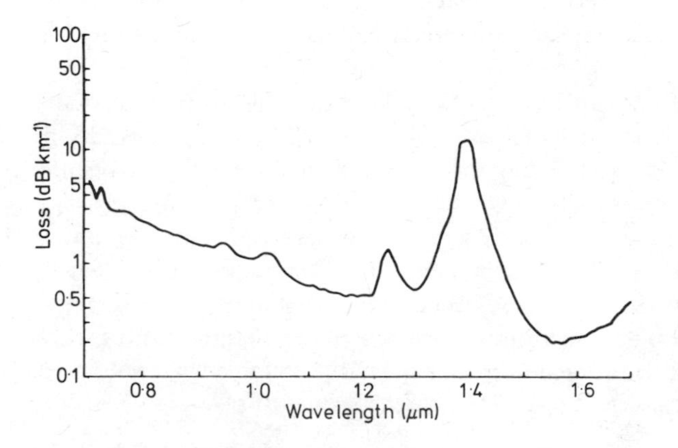

Figure 2. Loss spectrum of a single-mode fibre with SiO_2 cladding and a GeO_2 doped silica core of 9·4 μm diameter. (From Miya *et al* (1979).)

1978). Recently the existence of this zero-pulse-dispersion point has been confirmed by pulse-delay measurements using wavelength-tunable radiation generated in a fibre via the stimulated Raman effect (Cohen and Lin 1977, Lin *et al* 1978), and by the transmission of picosecond pulses over kilometre-lengths of single-mode fibre without observable pulse broadening (Bloom *et al* 1979). Apart from material dispersion there is additional pulse dispersion in multimode fibres caused by group velocity differences of the modes. This is absent in single-mode fibres, and figure 3 shows the dispersion characteristics of two single-mode fused-silica fibres with zero dispersion wavelengths near 1·35 and 1·5 μm.

There is still a considerable amount of work to be done to improve the properties of first-generation devices. Important issues here include degradation and the operating life of lasers, yield, stability and self-pulsations. While this work is proceeding, device research is also giving attention to second generation devices and beyond. One goal is to make available single-mode lasers for single-mode fibres. Another goal is to provide source and detector options for long-wavelength systems that can exploit the favourable fibre properties mentioned above. GaAlAs and Si are no longer suitable for these wavelengths, but considerable success with new materials such as InGaAsP has already been attained.

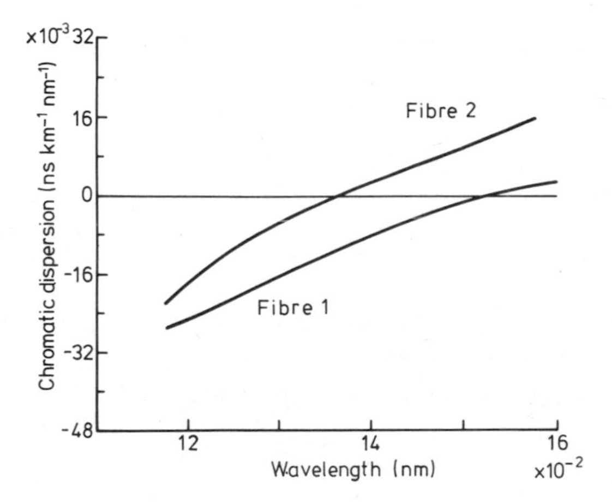

Figure 3. Dispersion spectrum of two GeO_2-doped single-mode silica fibres with core diameters of 5·2 μm (fibre 1) and 7 μm (fibre 2). (From Cohen *et al* (1979).)

In the following we will attempt to sketch the nature and trend of recent research on devices for lightwave communications. While the space and time available for this sketch are limited, the field has grown to such a large size that any selection of illustrations for this work must be subjective and incomplete, and the writer apologises in advance for all omissions.

The topics selected for discussion include the new semiconductor materials for optical devices, epitaxial growth methods used for device preparation, basic characteristics of new device materials, structures for single-mode lasers, laser degradation, self-pulsations in lasers and new detectors. We will also mention devices that offer possibilities for wavelength multiplexing and switching. These are GRIN-rod devices suitable for multimode fibre systems and integrated optics devices compatible with single mode fibres.

2. Semiconductor materials

The semiconductor lasers of interest for lightwave communications require the preparation of heterostructure crystal layers which are of different band gap but are matched in their lattice constants. Heterostructures are also used for LEDs and some detector devices. Today, by far the best developed materials system is the $Al_xGa_{1-x}As/GaAs$ system, which is providing lasers and LEDs for present lightwave communication systems (for recent reviews see Kressel and Butler 1977, Casey and Panish 1978). The band gap in this material can be tailored to span the wavelength range 0·8–0·9 μm by changing the (AlGa) composition.

In the search for materials suitable for longer wavelengths researchers have investigated a considerable variety of III–V compounds compatible with the substrate materials GaAs, InP and GaSb. A severe but important materials test was the demonstration of continuous laser operation at room temperature. The break into the long-wavelength range happened in 1976, when three materials passed that test (Nahory *et al* 1976, Hsieh *et al* 1976, Nuese *et al* 1976). The first was the $GaAs_{1-x}Sb_x/Al_yGa_{1-y}As_{1-x}Sb_x$ heterostructure system which was graded in lattice spacing to GaAs substrates (Nahory *et al* 1976). CW laser operation was obtained at 1·0 μm and later extended to 1·06 μm. The next was the quaternary system of $In_{1-x}Ga_xAs_yP_{1-y}/InP$ lattice-matched to InP substrates (Hsieh *et al* 1976) and initially operating at 1·1 μm. The third CW room temperature laser operation was demonstrated in $In_xGa_{1-x}As/In_yGa_{1-y}P$ heterostructures graded in lattice spacing to GaAs substrates (Nuese *et al* 1976), and operating at wavelengths of 1·06 and 1·12 μm.

At present the focus of research is on the (InGa) (AsP) quaternary system which offers good lattice match (by holding the compositions in the proportion $y \simeq 2{\cdot}2x$). The band gap in this system can be tailored for operation from about 1·0 to 1·7 μm. In the long-wavelength limit we have $y = 1$ and are dealing with the ternary system $In_{0{\cdot}53}Ga_{0{\cdot}47}As/InP$, which is also a desirable detector material.

Another promising detector material is the ternary $Al_yGa_{1-y}As_{1-x}Sb_x$, which can be lattice matched to GaSb substrates (Law *et al* 1978). In fact, GaSb appears to be a better quality substrate at present compared to InP. However, much effort is devoted to the improvement of InP, and this should soon bear fruit.

3. Methods of epitaxial growth

There are essentially three different methods which are used for the preparation of epitaxial heterostructures: (1) liquid phase epitaxy (LPE), (2) vapour phase epitaxy (VPE), and (3) molecular beam epitaxy (MBE).

LPE is the simplest, least expensive and most commonly used technique. Here, the epitaxial layers are grown from solutions with growth rates of about 100–1000 nm min^{-1}. The first CW room-temperature lasers in both the (AlGa)As and the (InGa) (AsP) systems were grown by LPE. LPE-grown lasers in both materials systems can now be made with thresholds better than 1 kA cm^{-2}.

The VPE and MBE methods promise better uniformity and yield. In the VPE method the epitaxial material is introduced in gaseous form, which offers possibilities for smooth grading of layers and flexibility for the introduction of dopants. Growth rates are of the order of 100–500 nm min^{-1}. The most successful VPE technique for the growth of

(AlGa)As heterostructures has been the metal–organic chemical vapour deposition (MO–CVD) technique (Dupuis *et al* 1977). VPE-grown (AlGa)As lasers have been demonstrated in CW room-temperature operation and with thresholds as good as or better than the best LPE material. Recently, VPE-grown (InGa) (AsP) lasers have also been demonstrated to be capable of CW room-temperature operation, and thresholds as low as $2\,kA\,cm^{-2}$ were achieved (Olsen *et al* 1979).

At present, MBE is probably the most complex and most expensive of the three techniques, but it offers the broadest range of materials possibilities and thickness control on a 10-Å scale. MBE is essentially an evaporation technique and requires a high vacuum of about 10^{-10} Torr. MBE growth rates are about $5–50\,nm\,min^{-1}$.

MBE-grown (AlGa)As lasers have been shown to be capable of CW operation at room-temperature (Cho *et al* 1976), and recently thresholds as low as $800\,A\,cm^{-2}$ have been achieved (Tsang 1979). Figure 4 shows a comparison of thresholds achieved in (AlGa)As lasers by the three different methods of epitaxial growth. No CW room-temperature lasers have as yet been made with MBE-grown (InGa) (AsP) material. However, a first step has been made with the achievement of $3\,kA\,cm^{-2}$ room-temperature thresholds in MBE-grown (InGa)As lasers operating pulsed at 1·65 μm (Miller *et al* 1978).

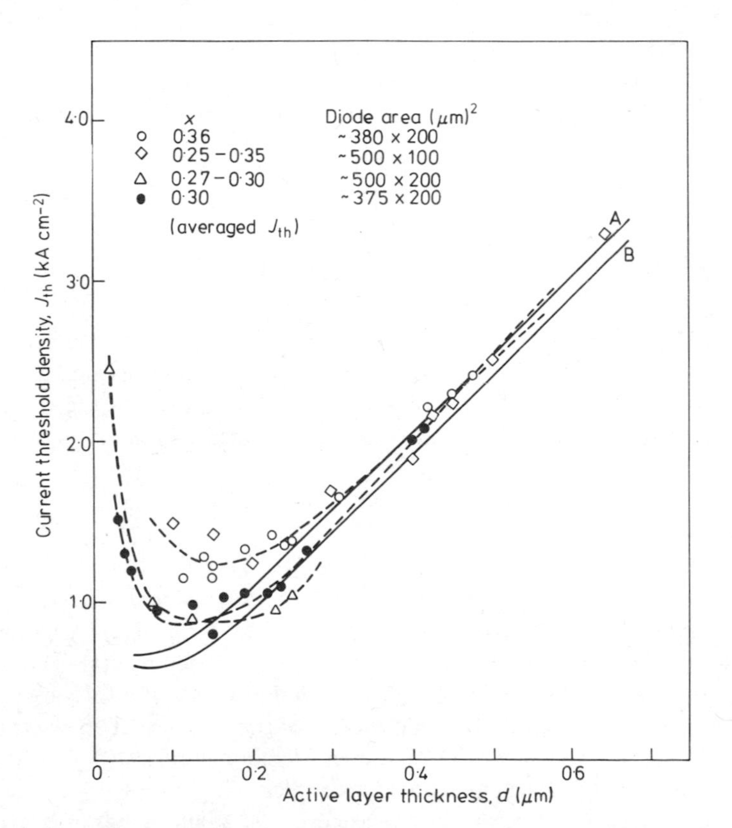

Figure 4. Comparison of threshold currents achieved in (GaAl)As heterostructure lasers as a function of active layer thickness and method of epitaxial growth. LPE data are indicated by open circles and diamonds, MO–CVD data by triangles, and MBE data by dots. $J_{th}/d = 5160\,A\,cm^{-2}\,\mu m$ for curve A and $4800\,A\,cm^{-2}\,\mu m$ for curve B. (After Tsang (1979).)

4. Characteristics of new device materials

Experimental efforts are under way to determine the characteristics of the new III–V materials. This is leading to a better understanding of their applicability to optical devices as well as to other electronic devices such as high-speed FETs. Among the results already accomplished is the accurate measurement of the band gap as a function of composition for (InGa) (AsP) material lattice-matched to InP (Nahory *et al* 1978). This is shown in figure 5.

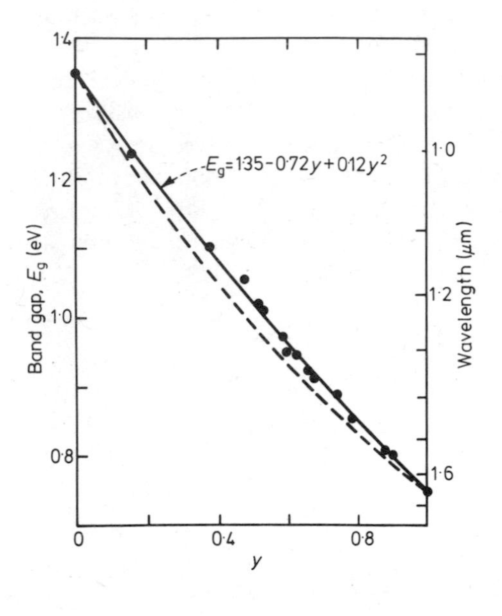

Figure 5. Band gap energy E_g as a function of the As content y for $In_{1-x}Ga_xAs_yP_{1-y}$ lattice-matched to InP. The full curve is the best fit to the measured data, the broken curve is an earlier prediction. (From Nahory *et al* (1978).)

The temperature dependence of the laser threshold gives indications for the behaviour of lasers to be expected at elevated temperatures, and provides clues on mechanism of carrier loss that might be present in the structure. The dependence for the threshold current density $J_{th}(T)$ is expected to be approximately exponential (Kressel and Butler 1977, Casey and Panish 1978), i.e.

$$J_{th} \propto \exp(T/T_0),$$

with a characteristic temperature T_0 describing the quality of the material. For (AlGa)As, T_0 is usually in the range 120–165 K (Casey and Panish 1978). For (InGa) (AsP) heterostructures, T_0 has been measured to be about 70 K near room temperature, both for LPE and VPE material (Olsen *et al* 1979, Horikoshi and Furukawa 1979, Nahory *et al* 1979). Measurements of T_0 have given some indications of the presence of an electrically active recombination centre in (InGa) (AsP) material (Nahory *et al* 1979). This centre affects device operation and should be eliminated if possible.

Another materials property that can influence the operational properties of lasers and other devices is due to the presence of an auxiliary indirect band gap. Electrons are lost from the upper level laser population by intraband scattering from the direct valley to the auxiliary valley. This scattering is a function of the energy difference ΔE between the direct and the indirect gap. The following is a table of ΔE in meV for a selection of III–V compounds of interest:

GaAs	300
InP	400
GaSb	80–90
InGaAs	800

The low ΔE-value for GaSb may present a serious hurdle for the accomplishment of CW room-temperature laser operation in that material.

As the new compounds may also be of interest for high-speed electronic devices, measurements are under way to determine their mobility and similar characteristics. On the basis of such measurements the mobility of low impurity (InGa) (AsP) lattice matched to InP has been projected to increase monotonically as a function of the As concentration from a value of 4700 $cm^2\,V^{-1}\,s^{-1}$ for InP to a value of 15 000 $cm^2\,V^{-1}\,s^{-1}$ for (InGa)As (Leheny *et al* 1980).

5. Single-mode laser structures

For efficient coupling to single-mode fibres one requires junction lasers that operate in a stable fundamental transverse mode. Frequently single transverse mode operation brings additional benefits such as kink-free light output against current characteristics. A considerable variety of single-mode structures providing lateral optical confinement has been investigated for GaAs lasers. Among the more successful ones are the buried heterostructure (BH) laser (Tsukada 1974), the transverse junction stripe (TJS) laser (Namizaki *et al* 1974), the channelled-substrate planar (CSP) laser (Aiki *et al* 1978) and the strip-buried heterostructure (SBH) laser (Tsang *et al* 1978). Cross sections of four laser structures are sketched in figure 6. The SBH lasers use a thin and narrow active strip of GaAs

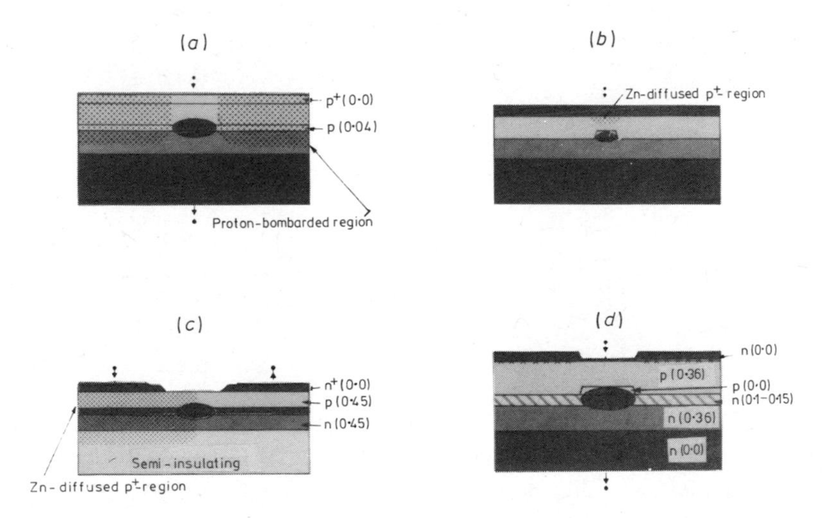

Figure 6. Cross section of: (*a*) proton-bombarded stripe geometry laser, (*b*) the BH laser, (*c*) the TJS laser, and (*d*) the SBH laser; $n(x) = n - Al_xGa_{1-x}As$. (Courtesy of W T Tsang.)

which acts as loading for an adjacent (AlGa)As layer to effect lateral mode guiding. Recent results with SBH lasers using 5–10 μm strips have exhibited about 85 mA CW thresholds, stable single transverse modes up to 9x threshold, excellent linearity up to 35 mW of CW power output per mirror, and narrow beam divergence (Tsang and Logan 1979). Recent results with TJS lasers on semi-insulating GaAs substrates are also very encouraging. In these structures the junction is made perpendicular to the epitaxial layers. The lasers are showing 12 mA thresholds, up to 15 mW CW output at room temperature, single transverse and single longitudinal modes, and long life (Kumabe *et al* 1979).

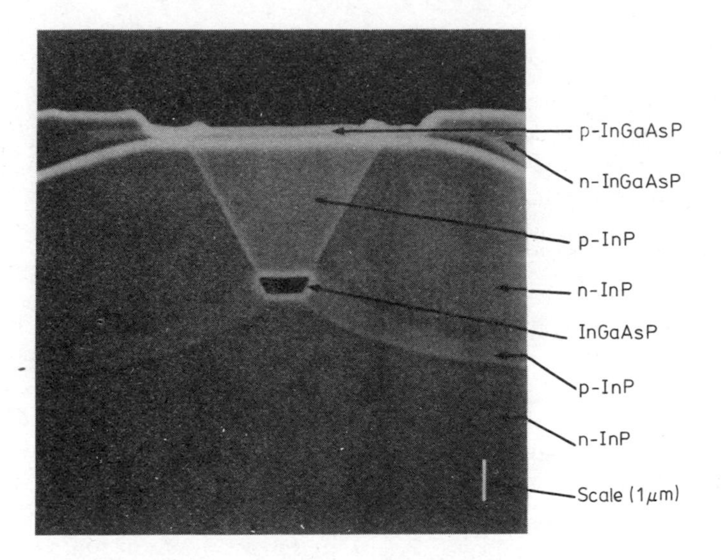

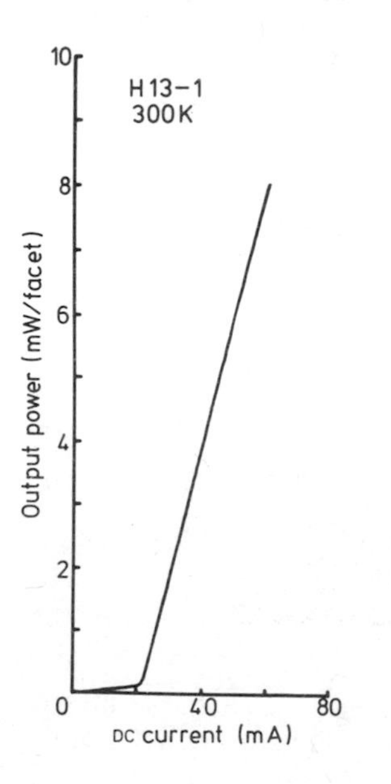

Figure 8. Output characteristics of the single-mode (InGa) (AsP) laser reported by Hirao *et al* (1979).

A recent achievement is the single-mode operation in (InGa) (AsP) lasers at a wavelength of 1·3 μm (Hirao *et al* 1979). The structures used in this work were BH lasers of 1–2 μm stripe width having CW room-temperature thresholds of 22 mA, CW output powers of well over 5 mW and demonstrated life of over 1000 h without deterioration. Figures 7 and 8 show the structure and the light-output against current characteristics of these lasers.

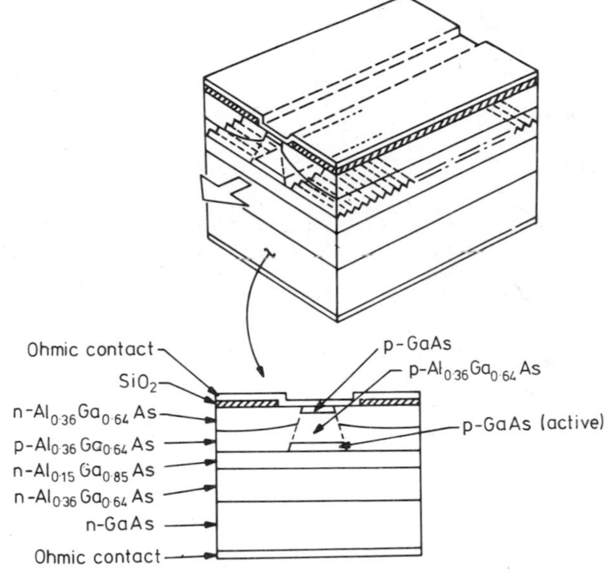

Figure 9. Sketch of the structure of a DFB laser with lateral-evanescent-field distributed feedback. (After Tsang *et al* (1979).

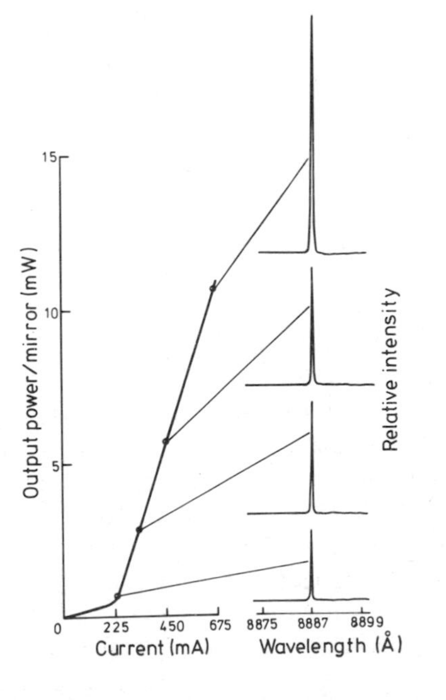

Figure 10. Light-output against current characteristics and associated output spectra of the DFB laser sketched in figure 9. (After Tsang *et al* (1979).)

Distributed feedback lasers (DFB) are structures that promise stable output frequencies and single longitudinal mode operation well above threshold. These structures are made by introducing high-resolution (~2000 Å) corrugations into the heterostructure layers. The technology for the preparation of these structures is still rather difficult. However, room temperature CW operation in GaAs DFB lasers has been demonstrated (Nakamura *et al* 1975). Recently a device was demonstrated that ensures single transverse modes by use of an SBH structure and provides DFB by means of corrugations in the evanescent field near the strip (Tsang *et al* 1979). Single longitudinal modes and stable output frequencies were obtained up to 3x threshold. Figures 9 and 10 show the structure and the output characteristics of this DFB laser.

6. Laser degradation

The operating life of CW junction lasers is, clearly, a vital parameter for systems applications. At this point it should be recalled that one deals with current densities of the order of $1\,\mathrm{kA\,cm^{-2}}$ in these devices. In almost 10 years of careful work, the projected mean life of GaAs junction lasers for CW operation at room temperature has been improved from a few hours to the order of one million hours as illustrated in figure 11. Life projec-

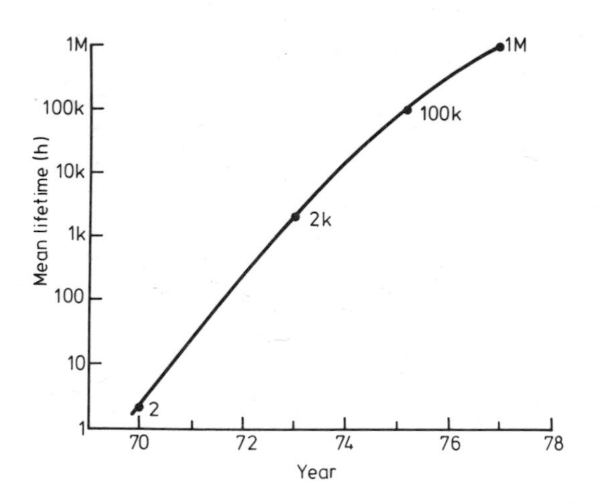

Figure 11. Progress in improvement of the reliability of (AlGa)As injection lasers with room-temperature CW operation. (From Li (1978).)

tions of this long duration are made on the basis of temperature-accelerated aging tests in the laboratory. Projections of 10^6 h mean lifetimes have now been reported for a variety of GaAs laser structures (Hartman *et al* 1977, Thompson 1979, Ettenberg 1979, Ishikawa *et al* 1979), including a single-mode TJS laser (Takamiya *et al* 1979).

(InGa) (AsP) junction lasers have been reported to continue operating after 10^4 h of CW operation. For a new material, this is extremely encouraging, but much careful work has yet to be done, and the study of relevant degradation mechanisms is just beginning.

The degradation mechanisms occurring in GaAs lasers are reviewed in recent text books (Kressel and Butler 1977, Casey and Panish 1978). They include gradual bulk degradation, facet damage and contact metallisation failure. It has been argued that bulk laser degradation in (InGa) (AsP) should be considerably reduced as the photon energy $h\nu$ is smaller than in GaAs, and degradation times should be proportional to $\exp(E_{\mathrm{act}}/h\nu)$, where E_{act} is the activation energy of the degradation mechanism. However, this argument holds only if there is no new degradation mechanism in the new materials. Indeed, the first studies of optical degradation in (InGa) (AsP) indicate that there is degradation due to a slip-and-glide mechanism that is not normally observed in GaAs (Johnston *et al* 1978, Mahajan *et al* 1979). These findings are based on spatially resolved photoluminescence studies and on transmission electron microscopy. They report dark line defects in the ⟨110⟩ direction due to the slip-and-glide mechanism. Dark lines in the ⟨100⟩ direction due to the climb mechanism, which is dominant in GaAs, are virtually absent in (InGa) (AsP).

7. Laser pulsations

The occurrence of self-pulsations in junction lasers is one of the remaining problems in (AlGa)As double-heterostructure lasers. These pulsations occur in various laser structures, and under various operating conditions, and they can appear after several hundred hours of aging (Paoli 1977). Both transient as well as continuous self-sustained pulsations have been observed in the frequency range 0·2–3 GHz. The pulsation frequency is related to the spontaneous recombination lifetime of the carriers of about 2 ns, and to the photon lifetime in the laser cavity of about 10 ps. As they interfere with the high-data-rate modulation of lasers, there are efforts to understand, control and eliminate these pulsations. There are many possible mechanisms for laser pulsations, including filament formation, temperature-induced changes and quantum shot-noise effects (for a recent review see Kressel and Butler 1977). Phenomenologically, the pulsations in single transverse mode lasers are caused by saturable absorption and/or saturable gain. However, the detailed origin of the responsible saturation effects is still unverified. Recent models trace the pulsations to absorbing dark line defects, to regions of carrier depletion (Dixon and Joyce 1979) and, possibly, to deep-level traps (Copeland 1978). Recent experiments with strip-buried heterostructure lasers find a strong correlation between the occurrence of pulsations and the presence of visible defects (Hartman *et al* 1979).

A better understanding of the nonlinear absorption and nonlinear gain mechanisms responsible for self-pulsations may not only aid in their elimination, but it can also provide the basis for the controlled generation of pulses in junction lasers (see e.g. Lee and Roldan 1970). In fact, some of these mechanisms have been used in recent work on the

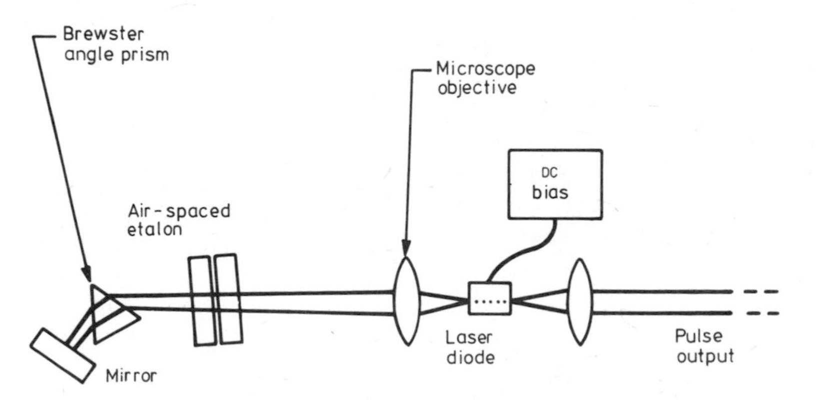

Figure 12. External cavity arrangement used for the generation of picosecond pulses from semiconductor junction lasers. (From Ippen *et al* (1979).)

generation of picosecond pulses in semiconductor lasers. Continuous pulse streams with a 3 GHz repetition rate and pulse duration of 20 ps have been obtained in GaAlAs lasers using an external resonator and active modulation with microwaves at 3 GHz (Ho *et al* 1978). Pulsewidths as short as 6 ps have been obtained in very recent experiments using SBH GaAlAs lasers and passive modelocking in external cavities with band-limiting elements (Ippen *et al* 1979). Figure 12 shows the cavity arrangement used in these experiments, and figure 13 shows a correlation function obtained by second-harmonic generation (SHG) and used for the determination of pulse width.

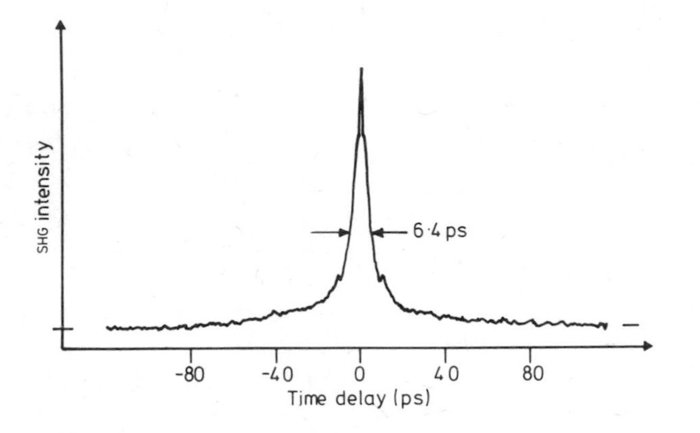

Figure 13. SHG correlation function of semiconductor laser output indicating pulsewidth of about 6 ps. (From Ippen *et al* (1979).)

8. Photodetectors

The detectors required for lightwave communications are room-temperature devices of high sensitivity and fast response, that can detect the attenuated digital optical signal

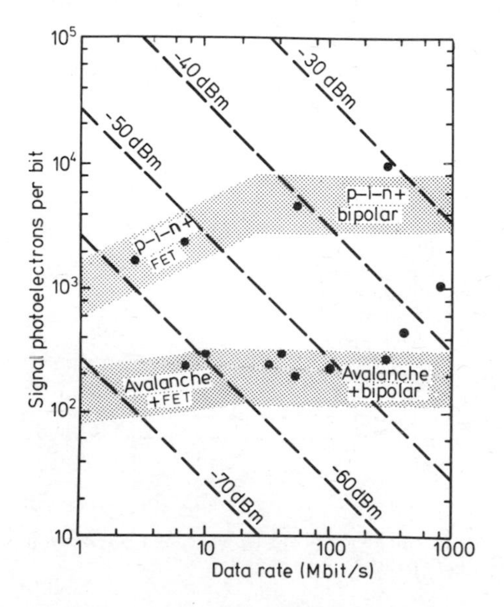

Figure 14. Calculated sensitivity (in photoelectrons per bit) of various optical receivers as a function of bit rate. The full circles represent experimental results achieved to date. (From Li (1978).)

with error rates better than 10^{-9} (for recent reviews see Melchior 1977, Melchior *et al* 1978). First-generation systems use Si p–i–n photodetectors for lower bit rates and Si avalanche photodetectors (APDs) for higher bit rates. This is illustrated in figure 14. Si p–i–n detectors are followed by low-noise amplifiers such as Si FETs. The internal gain in APDs is optimised to gain values ranging from about 10 to 100 in the presence of excess noise due to the avalanche multiplication process. At bit rates above 10 Mbit/s the APD receivers offer a sensitivity improvement of about 15 dB as compared to Si p–i–n receivers, for a typical overall sensitivity of 50 dBm (see e.g. Li 1978). However, APDs require relatively high voltages (100–400 V) and temperature compensation circuits to stabilise the gain. There is, therefore, continuing interest in improving p–i–n–FET combinations, and the recent availability of low-noise GaAs FETs has widened the applicability of p–i–n–FETs to higher speeds (see e.g. Hata *et al* 1977, Smith *et al* 1978).

Si detectors are no longer useful at wavelengths longer than 1·1 μm, and device research is busy trying to provide detectors for the long-wavelength range (for a recent review see Law *et al* 1979). Ge is a well developed material that is a good candidate for this purpose. However, Ge is an indirect gap material with a slow fall-off in response towards longer wavelengths and with relatively large dark currents. The search for better materials is focusing on InGaAs, InGaAsP and GaAsSb, which are direct gap materials and are sensitive up to about 1·7 μm. Figure 15 illustrates the spectral response of these materials and

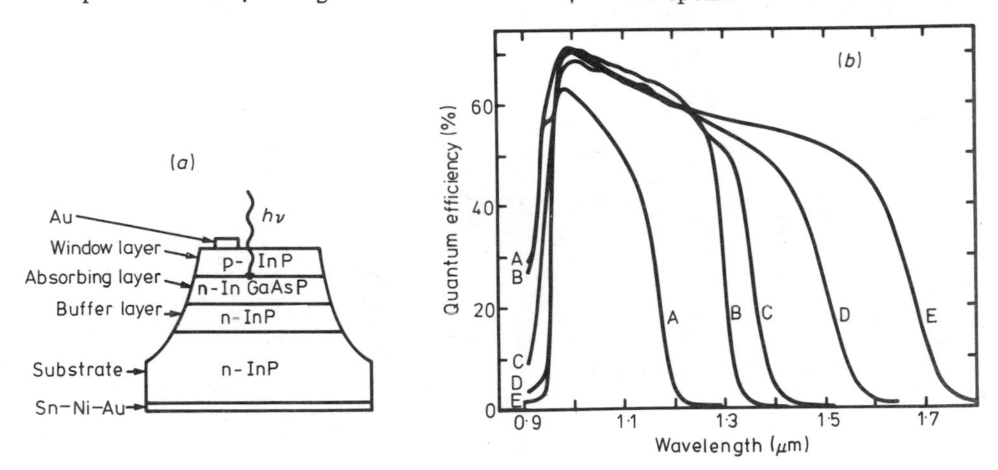

Figure 15. (*a*) Structure and (*b*) external quantum efficiency of five $In_{1-x}Ga_xAs_yP_{1-y}$ photo-detectors with different composition. The As content y was: A, 0·47; B, 0·61; C, 0·66; D, 0·88; and E, 1·0; respectively. (After Washington *et al* (1978).)

their sharp cut-off towards longer wavelengths, which can be moved by changing the material's composition (see e.g. Washington *et al* 1978). This is a very desirable feature as leakage currents tend to increase exponentially with the cut-off wavelength due to the smaller band gaps. For p–i–n–FET applications of these materials the requirements include those for low leakage current, low capacitance, and high breakdown voltage. Some encouraging results have already been reported where these parameters lie in range of 10 nA, 1 pF and 100 V (see e.g. Burrus *et al* 1979, Leheny *et al* 1979), but there are also some unexplained phenomena such as an increase in dark current with reverse-bias voltage.

The excess noise due to avalanche multiplication in APDs increases as the ionisation rates for electrons and holes become more equal. In Si these ionisation rates differ by more than an order of magnitude, but in Ge, InGaAs, InGaAsP and GaAsSb they appear to be almost the same. Avalanche gain in the new materials has been observed in several laboratories (for reviews see e.g. Melchior 1977, Law *et al* 1979), but the search for low-noise APD materials still continues. An impulse in this direction may be progress in the understanding of the connection between impact ionisation and basic materials properties such as band structure and crystal orientation (see e.g. Pearsall *et al* 1977).

The sharp cut-off properties of the new materials are also opening up new device possibilities. A recent example for this is a wavelength demultiplexing (InGa) (AsP)

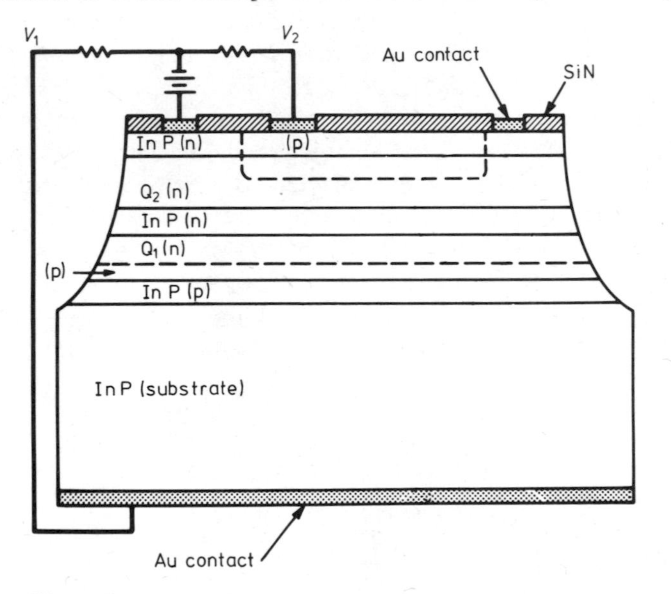

Figure 16. Sketch of the structure of a demultiplexing diode. The two (InGa) (AsP) layers are labelled Q_1 and Q_3. (From Campbell *et al* (1979).)

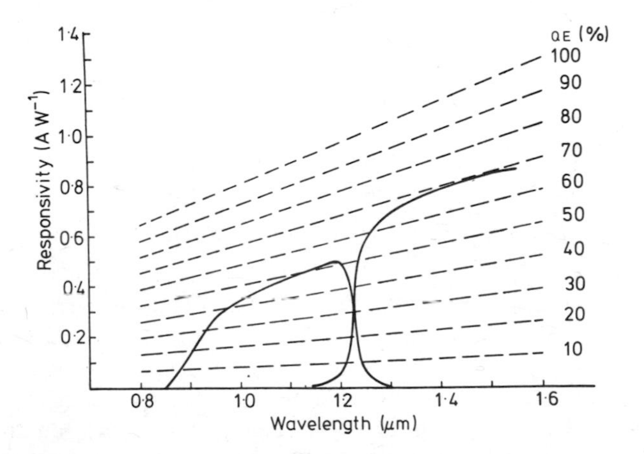

Figure 17. Photoresponse of the two channels of a demultiplexing diode. (Courtesy of T P Lee.)

photodiode, that detects and demultiplexes two wavelength bands simultaneously (Campbell *et al* 1979). The structure of such a device is sketched in figure 16 and its photoresponse is shown in figure 17.

9. GRIN-rod devices

GRIN-rods are glass rods with a parabolically graded refractive index profile similar to that used in graded index fibres. GRIN-rods act like lenses and can be used, in combination with optical elements such as gratings or semitransparent mirrors, to assemble very compact, rugged and stable devices for the manipulation and processing of optical signals. A variety of devices has been proposed and demonstrated, and most of them are compatible with first-generation multimode fibres, however some applications to single-mode fibre systems appear possible. Examples for GRIN-rod devices are attenuators, directional couples and switches (see e.g. Doi *et al* 1978, Kobayashi *et al* 1979), as well as multiposition switches (Tomlinson *et al* 1979) and wavelength multiplexers (Tomlinson and Aumiller 1977). A sketch of a typical wavelength multiplexer is shown in figure 18, and its demultiplexing filter characteristics are given in figure 19 for the case of two channels.

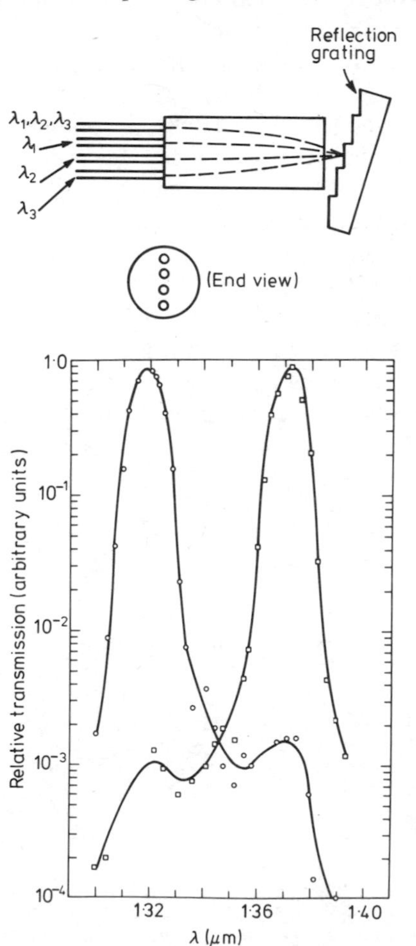

Figure 18. Sketch of a wavelength multiplexer consisting of a GRIN lens and a grating. Only three input fibres are shown. (After Tomlinson (1980).)

Figure 19. Filter response of a GRIN-rod demultiplexer in the near IR. The responses of two channels are shown. (After Tomlinson (1980).)

The latter provides an illustrative example for GRIN-rod devices. Such a multiplexer device might consist of five input and output fibres, a GRIN-rod lens and a blazed grating, assembled in a compact package 1–2 cm long and 2–4 mm in diameter. It would be capable of multiplexing or demultiplexing four optical channels of different wavelength, with a channel separation of 30 nm and crosstalk better than −30 dB.

10. Integrated optics devices

In integrated optics, dielectric waveguides are used on planar substrates to confine the light to small cross sections over relatively long lengths (for recent reviews see e.g. Tien 1977, Kogelnik 1978). One aims for miniaturised devices of improved reliability and stability, of lower power consumption and lower drive voltages. Devices of interest are couplers, junctions, directional couplers, filters, wavelength multiplexers and demultiplexers, modulators, switches, lasers and detectors. Apart from providing new or improved versions of these devices, a hope of integrated optics is to combine two or more devices on a single chip. Most integrated optics devices and circuits are single-mode devices which are compatible with single-mode fibres. Integrated optics may provide future lightwave communication systems with such sophisticated options as single-mode wavelength multiplexing circuits or electro-optic switching networks. Research in this field is broad and varied, but two examples will have to suffice here to illustrate its nature. Both devices are based on directional couplers fabricated in the electro-optic material $LiNbO_3$. The coupler consists of two waveguides, about 3 μm wide, which approach each other to a distance of about 3 μm over an interaction length of about 1 mm–1 cm. The guides are prepared by diffusing Ti into the substrate. Metal electrodes on the surface of the crystal allow the application of an electric field. This induces refractive index changes via the electro-optic effect and allows control of the coupler. Split electrodes and applied voltages of reversed polarity ensure that complete crossover of the light from one guide to the other can be achieved by an electrical adjustment (Kogelnik and Schmidt 1976). This is called the (alternating) $\Delta\beta$ coupler configuration and has been used in a variety of devices. One of them is a switch and amplitude modulator that includes six sections of alternating $\Delta\beta$, and can operate at data rates in excess of

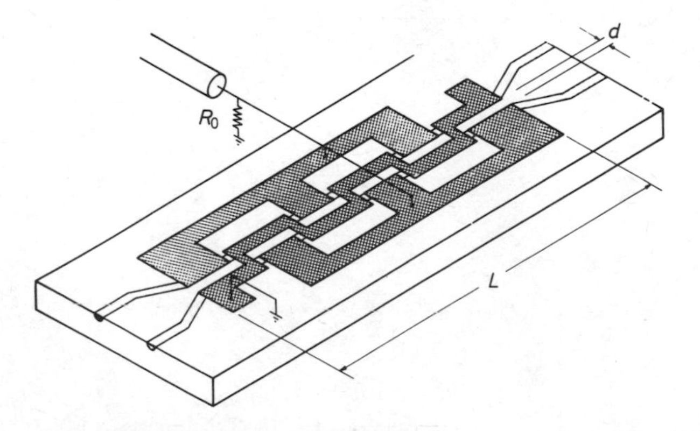

Figure 20. Sketch of alternating $\Delta\beta$ coupler consisting of a Ti diffused directional coupler of interaction length L and guide separation d. The electrodes (shown hatched) provide for six polarity reversals. (After Schmidt and Cross (1978).)

100 Mbit/s with drive voltages as low as 3 V (Schmidt and Cross 1978). The structure of this device is sketched in figure 20. The applied voltage switches the light from one output guide to the other. The detailed switching characteristics are shown in figure 21. The second example is a tunable filter device which is very similar in structure to the first. The difference is that the two waveguides are non-identical and have intersecting dispersion characteristics (Alferness and Schmidt 1978). To achieve this, the two waveguides are fabricated to different widths (1·5 and 3 μm) with different effective refractive index. The device has a measured filter bandwidth of 20 nm and is electrically tunable at a rate

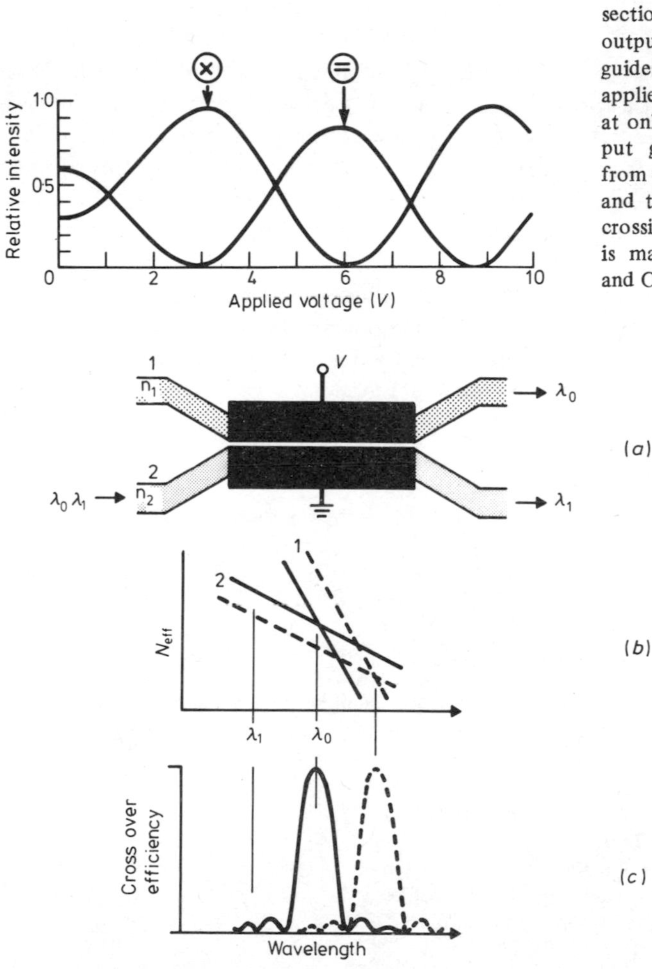

Figure 21. Switching characteristics of a $\Delta\beta$ coupler with six sections of alternating $\Delta\beta$. Relative output power in the two output guides is shown as a function of applied voltage for light injection at only one input guide. The output guide fed straight through from the input guide is marked = , and the ouput guide fed by light crossing over from the input guide is marked × . (After Schmidt and Cross (1978).)

Figure 22. Structure, dispersion and filter characteristics of a tunable directional coupler filter. The device structure (*a*) consists of a directional coupler consisting of two different guides marked 1 and 2 and electrodes for the application of a voltage (electrode split is not shown). The effective index N_{eff} of the two guides is shown as a function of wavelength λ indicating crossing from one guide. The filter response of the light crossing from one guide to the other is shown in (*c*). (After Alferness and Schmidt (1978).)

of 11 nm V^{-1}. The structure and characteristics of this directional coupler filter are illustrated in figure 22.

Finally, we should mention two examples, where several similar devices have been integrated on a single chip in the research laboratory. The first example is an experimental 4 x 4 optical switching network made by integrating five $\Delta\beta$ couplers on a $LiNbO_3$ substrate (Schmidt and Buhl 1976). The other example is an experimental wavelength multiplexing chip consisting of six DFB lasers operating at different wavelengths with a junction circuit combining the six outputs into one guide on a GaAs substrate (Aiki *et al* 1976).

11. Conclusions

Coupled with the rapid emergence of lightwave communications, there has been a growing research and development effort on optical devices. Device research in this field has seen considerable success, but many exciting challenges remain for solid-state, materials, device and systems specialists. Research interest is expanding from devices compatible with multimode fibres to those compatible with single-mode fibres, and from devices operating in the 0·8 μm region to devices capable of operating in the long-wavelength region near 1·3 and 1·6 μm.

Because of space and time limitations, the illustrations given in this paper of the diverse trends in device research had to be sketchy, incomplete and subjective. The many necessary omissions include detailed descriptions of several device aspects important in applications such as details on LEDs and the direct modulation of lasers.

References

Aiki K, Nakamura M, Kuroda T, Umeda J, Ito R, Chimone N and Maeda M 1978 *IEEE J. Quantum Electron.* **14** 89

Aiki K, Nakamura M and Umeda J 1976 *Appl. Phys. Lett.* **29** 506

Alferness R C and Schmidt R V 1978 *Appl. Phys. Lett.* **33** 161

Bell Syst. Tech. J. 1978 **57** 1717 (see collection of articles)

Bloom D M, Mollenauer L F, Lin C, Taylor D W and DelGaudio A M 1979 *Opt. Lett.* **4** to be published

Burrus C A, Dentai A G and Lee T P 1979 *Electron. Lett.* **15** 655

Campbell J C, Lee T P, Dentai A G and Burrus C A 1979 *Appl. Phys. Lett.* **34** 401

Casey H C Jr and Panish M B 1978 *Heterostructure Lasers* (New York: Academic Press)

Cho A Y, Dixon R W, Casey H C Jr and Hartman R L 1976 *Appl. Phys. Lett.* **28** 501

Cohen L G and Lin C 1977 *Appl. Opt.* **16** 3136

Cohen L G, Lin C and French W G 1979 *Electron. Lett.* **15** 334

Copeland J A 1978 *Electron. Lett.* **14** 809

Dixon R W and Joyce W B 1979 *IEEE J. Quantum Electron.* **16** 470

Doi K, Nonaka S, Yunki T and Takahashi M 1978 *NEC Res. Dev.* **50** 17

Dupuis R D and Dapkus P D 1977 *Appl. Phys. Lett.* **31** 8391

Ettenberg M 1979 *J. Appl. Phys.* **50** 1195

Hartman R L, Logan R A, Koszi L A and Tsang W T 1979 *J. Appl. Phys.* to be published

Hartman R L, Schumaker N E and Dixon R W 1977 *Appl. Phys. Lett.* **31** 756

Hata S, Kajiyama K and Mizushima Y 1977 *Electron. Lett.* **13** 668

Hirao M, Doi A, Tsuji S, Nakamura M and Aiki K 1979 *Appl. Phys. Lett.* to be published

Ho P T, Glasser L A, Ippen E P and Haus H A 1978 *Appl. Phys. Lett.* **33** 241

Horikoshi Y and Furukawa Y 1979 *Jap. J. Appl. Phys.* **18** 809

Hsieh J J, Rossi J A and Donnelly J P 1976 *Appl. Phys. Lett.* **28** 709

Ippen E P, Eilenberger D J and Dixon R W 1979 to be published
Ishikawa H, Fujiwara T, Fujiwara K, Morimoto M and Takusagawa M 1979 *J. Appl. Phys.* **50** 2518
Jacobs I and Miller S E 1977 *IEEE Spectrum* **14** 33
Johnston W D Jr, Epps G Y, Nahory R E and Pollack M A 1978 *Appl. Phys. Lett.* **33** 992
Kobayashi K, Ishikawa R, Minemura K and Sugimoto S 1979 *Fiber and Integrated Optics* **2** 1
Kogelnik H 1978 *Fiber and Integrated Optics* **1** 227
Kogelnik H and Schmidt R V 1976 *IEEE J. Quantum Electron.* **12** 396
Kressel H and Butler J K 1977 *Semiconductor Lasers and Heterojunction LEDs* (New York: Academic Press)
Kumabe H, Tanaka T, Namizaki H, Takamiya S, Ishi M and Susaki W 1979 *Jap. J. Appl. Phys.* **18** *Suppl.* 1 371
Law H D, Nakano K and Tomasetta L R 1979 *IEEE J. Quantum Electron.* **15** 549
Law H D, Nakano N, Tomasetta L R and Harris J S 1978 *Appl. Phys. Lett.* **33** 948
Lee T P and Roldan R H R 1970 *IEEE J. Quantum Electron.* **6** 339
Leheny R F, Ballman A A, DeWinter J C, Nahory R E and Pollack M A 1980 to be published
Leheny R F, Nahory R E and Pollack M A 1979 *Electron. Lett.* to be published
Li T 1978 *IEEE Trans. Commun.* **26** 946
Lin C, Cohen L G, French W G and Foertmeyer V A 1978 *Electron. Lett.* **14** 170
Mahajan S, Johnston W D Jr, Pollack M A and Nahory R E 1979 *Appl. Phys. Lett.* **34** 717
Melchior H 1977 *Phys. Today* **30** 32
Melchior H, Hartman A R, Schinke D P and Seidel T E 1978 *Bell Syst. Tech. J.* **57** 1791
Miller B I, McFee J H, Martin R J and Tien P K 1978 *Appl. Phys. Lett.* **33** 44
Miya T, Terunuma Y, Hosaka T and Miyashita T 1979 *Electron. Lett..* **15** 106
Nahory R E, Pollack M A, Beebe E D, DeWinter J C and Dixon R W 1976 *Appl. Phys. Lett.* **28** 19
Nahory R E, Pollack M A and DeWinter J C 1979 *Electron. Lett.* to be published
Nahory R E, Pollack M A, Johnston W D Jr and Barns R L 1978 *Appl. Phys. Lett.* **33** 659
Nakamura M, Aiki K, Umeda J and Yariv A 1975 *Appl. Phys. Lett.* **27** 403
Namizaki H, Kan H, Ishii M and Ito A 1974 *J. Appl. Phys.* **45** 2785
Nuese C J, Olsen G H, Ettenberg M, Garmon J J and Zamerowski T J 1976 *Appl. Phys. Lett.* **29** 807
Olsen G H, Nuese C J and Ettenberg M 1979 *Appl. Phys. Lett.* **34** 262
Paoli T L 1977 *IEEE J. Quantum Electron.* **13** 351
Pearsall T P, Nahory R E and Chelikowsky J R 1977 *Phys. Rev. Lett.* **39** 295
Schmidt R V and Buhl L L 1976 *Electron. Lett.* **12** 575
Schmidt R V and Cross P S 1978 *Opt. Lett.* **2** 45
Schwartz M I, Runstra W A, Mullins J H and Cook J S 1978 *Bell Syst. Tech. J.* **57** 1881
Smith D R, Hopper R C and Garrett I 1978 *Opt. Quantum Electron.* **10** 292
Takamiya S, Namizaki N, Susaki W and Shirahata K 1979 *Digest IEEE/OSA Conf. on Laser Engineering and Applications* p 51
Thompson A 1979 *IEEE J. Quantum Electron.* **15** 11
Tien P K 1977 *Rev. Mod. Phys.* **49** 361
Tomlinson W J 1980 *Appl. Opt.* **19** to be published
Tomlinson W J and Aumiller G D 1977 *Appl. Phys. Lett.* **31** 169
Tomlinson W J, Wagner R E, Strand A R and Dunn F A 1979 *Electron. Lett.* **15** 192
Tsang W T 1979 *Appl. Phys. Lett.* **34** 473
Tsang W T and Logan R A 1979 *IEEE J. Quantum Electron.* **15** 541
Tsang W T, Logan R A and Ilegems M 1978 *Appl. Phys. Lett.* **32** 311
Tsang W T, Logan R A and Johnson L F 1979 *Appl. Phys. Lett.* **34** 752
Tsukada T 1974 *J. Appl. Phys.* **45** 4899
Washington M A, Nahory R E, Pollack M A and Beebe E D 1978 *Appl. Phys. Lett.* **33** 854

Silicon carbide devices

E Pettenpaul†, W von Münch‡ and G Ziegler§

† Siemens AG, Balanstrasse 73, D-8000 München, Germany

‡ Institut für Halbleitertechnik, Breitscheidstrasse 2, D-7000 Stuttgart, Germany

§ Siemens AG, Werner-von-Siemens-Strasse 50, D-8520 Erlangen, Germany

Abstract. Large-gap semiconductors, like AlN, GaN, ZnS, SiC and diamond, are potentially useful materials for the realisation of light-emitting diodes for a wide spectral range, including the blue and UV regions. In addition, some of these materials are taken into consideration for the fabrication of high-power microwave devices and sensors for high-temperature operation. The knowledge of some physical properties of SiC (e.g. thermal conductivity, saturated drift velocity, breakdown field) has been extended during the past three years. A figure of merit, therefore, can be derived for some microwave devices based on SiC. These figures will be compared with the properties of conventional devices.

The techniques for the growth of SiC crystals and epitaxial layers are described in detail. Both p- and n-type substrates and epitaxial layers with doping levels in the $10^{15}-10^{19}$ cm^{-3} range have been prepared. The process steps for the generation of SiC MESFETs (Schottky-barrier field-effect transistors) with a gate-length of 1 μm are described and the results of static measurements on these devices are presented. It has been demonstrated that blue-emitting SiC diodes with reasonable light output can be produced by methods which are similar to those of present GaAs or GaP devices. The physics of the radiative recombination processes and the optimisation of the light output are discussed.

1. Introduction

Large-gap semiconductors, like AlN, GaN, ZnS, SiC and diamond, are potentially useful materials for the realisation of light-emitting diodes for a wide spectral range, including the blue and UV regions. In addition, some of these materials are taken into consideration for the fabrication of high-power microwave devices and sensors under extreme working conditions.

From the historical point of view SiC can be considered to be one of the oldest semiconductors, because the observation of electroluminescence was reported by Round in 1907 when he was working with SiC crystals. Moreover, crystallographers expressed scientific interest in SiC because it shows, like ZnS, CdS and KI, the phenomenon of polytypism, although large-area crystals from which devices can be fabricated are only the 6H (hexagonal) and to some extent the 15R (rhombohedral) structures.

The semiconductor properties of SiC are similar to silicon, although there are some essential differences in terms of its partly covalent character, a non-axial crystal structure, a large unit cell and a large band gap, all of which make the interpretation of the physical phenomena observed very complex.

Contrary to other large-gap semiconductors, SiC can easily be doped both p- and n-type; nevertheless difficult techniques had to be developed because of its extreme

0305-2346/80/0053-0021 $02.00

hardness and chemical inertness. This fact has restricted the work on it, and only a handful of laboratories have been engaged in SiC research for any length of time.

2. Single-crystal growth

The nature of the silicon–carbide binary system makes three crystal growth methods possible: (i) pyrolysis of volatile compounds of silicon and carbon; (ii) crystallisation from silicon or metal melt; and, (iii) sublimation.

Previous studies of vapour-phase reactions involving SiC were performed with emphasis on single-crystal growth of different polytypes (Kendall 1953, Berman and Ryan, 1971, Kobayashi *et al* 1975). The thermal reduction of silicon and carbon compounds makes it possible to guarantee an adequate control of growth conditions and polytype. Moreover, it is possible to grow very pure crystals. The disadvantage of this method, however, is the low growth rate, which has prevented the synthesis of single crystals of adequate size so far. In view of the substantial progress in the preparation of polycrystalline SiC with reasonable growth rate using silane compounds (von Münch and Pettenpaul 1978) the preparation of single crystals with methyl-trichlorsilane in the 1500–1600 °C temperature range by proper choice of the growth conditions may be successful.

Crystallisation from the stoichiometric silicon–carbon melt is a technological problem under ordinary conditions, because a temperature in excess of 2000 °C together with a pressure of perhaps 10^5 atm is needed. Therefore, working with either a carbon-saturated non-stoichiometric silicon melt or an adequate metal melt containing silicon and carbon may be a way. In the first case, crystallisation from a silicon-rich melt, the problem is to achieve appreciable solubility of carbon in silicon, a temperature of about 1700 °C is needed (Scace and Slack 1959) where evaporation of silicon becomes rapid and should be suppressed. It would be attractive to employ a graphite crucible with liquid silicon and to use the normal crystal-pulling technique, but the disadvantage here is that the crucible is dissolved. A SiC-coated crucible in connection with a method like the liquid-encapsulated Czochralski technique may be successfully applied to this material. It appears that the solubility of carbon in a (Cr or Fe)–silicon alloy is some times higher than in pure silicon at 1500–1700 °C (Halden 1960, Griffiths and Mlavski 1964, Knippenberg and Verspui 1966). A moderate growth rate was noted, but it seems that it is not a very promising method, either for high purity or for producing large crystals of high perfection.

The most successful method of growing SiC to date is the sublimation process. Single crystals of 2–3 cm in diameter are sometimes found in hollow cavities when preparing SiC grinding powder in the so-called Acheson (1892) process. In this process a mixture of silica and carbon with a few percent of sawdust and common salt (containing aluminium compounds) is heated in a trough-type electric furnace to about 2000 °C. Although they show a high degree of impurity, p-type crystals prepared in this way can be used as substrates for light-emitting diodes. Nevertheless, a systematic analysis of the semiconductor properties of SiC began with Lely's first demonstration of the sublimation technique in the laboratory in 1955.

The sublimation system which is used by the authors is shown in figures 1(*a*) and (*b*). Details of the apparatus and procedures are similar to those of Kapteyns and Knippenberg (1970), although special precautions were taken to optimise the vertical and horizontal

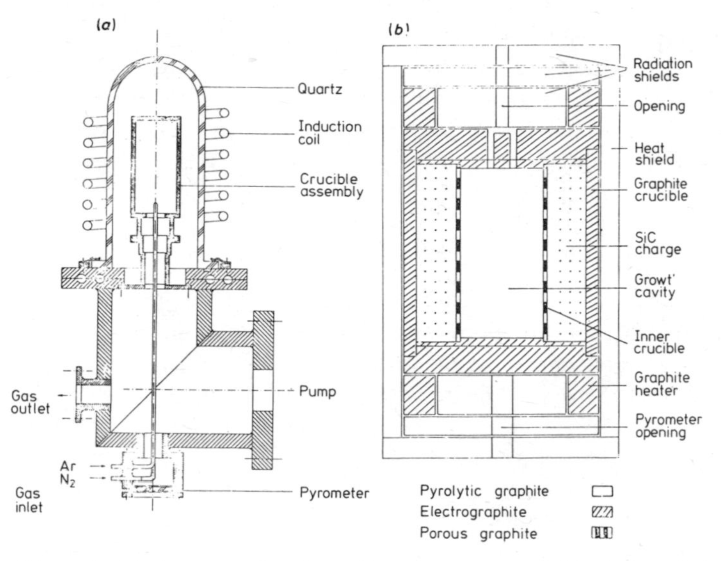

Figure 1. (*a*) Apparatus for the sublimation growth of SiC (Lely technique). (*b*) Crucible assembly for the growth of SiC crystals.

temperature gradients within the growth chamber by use of a suitable combination of graphite heat shields and additional induction heaters (figure 1*b*). Vertical and horizontal temperature gradients of 6 and $2\,^{\circ}C\,cm^{-1}$ were established at a growth temperature of $2550\,^{\circ}C$.

The doping level of SiC crystals grown by the Lely technique depends strongly on the impurity content of the starting material, the quality of the graphite parts and the purity of the argon atmosphere, as well as temperature and duration of degassing cycles.

In most cases, so far, commercial (green or black) grinding powder has been used as a feed material (Lely 1955, Knippenberg 1963, Potter and Sattele 1972), but the synthesis from the elements is generally less efficient in terms of crystal size and quality. With the grinding powder, however, large amounts of various impurities are introduced into the growth system. Partly compensated p-type material with a carrier concentration between 10^{15} and $10^{16}\,cm^{-3}$ can be obtained if 'green grit' abrasive is used as the starting material, provided that meticulous cleaning and degassing procedures are applied before and during the initial stages of the Lely process. Crystals of this type have been used as substrates for n-channel FETs.

SiC crystals with very low impurity content can be produced if the starting material is generated by pyrolysis of silane compounds (von Münch and Pettenpaul 1978). The carrier concentration of such crystals is n-type around $10^{16}\,cm^{-3}$ Nitrogen doping during the Lely process yields n-type crystals with an ionised donor concentration in the 5×10^{16}–$5\times10^{19}\,cm^{-3}$ range. The highly doped crystals would be required for the substrates of IMPATT diodes. The preparation of p-type substrate crystals was done by adding 'black' abrasive material or by the introduction of trimethyl-aluminium during pyrolytic decomposition of silane compounds for starting material. Low-resistivity substrates for LEDs can be prepared by adding Al_4C_3 to the charge of 'black' abrasive material or to the aluminium-doped CVD material after degassing cycles.

In summary, recent developments of the Lely technique have resulted in considerable improvements in yield and crystal quality. Both p- and n-type hexagonal platelettes with maximum dimensions of 10 mm and doping levels in the 5×10^{15}–$5 \times 10^{19}\,cm^{-3}$ range have been prepared. Material from this process is good enough for the fabrication of semiconductor devices although the lack of large-area single-crystal substrates is a serious restriction to large-scale production of SiC devices.

3. Junction fabrication

In SiC p–n junctions can be produced by diffusion, epitaxial growth and ion implantation. The diffusion of dopants, however, requires temperatures around 1900 °C. This temperature is considerably above the melting point of SiO_2 and no commercially attractive alternative diffusion masking material has yet been found. Moreover, all known impurities diffuse very slowly at temperatures below 1800 °C and yield very shallow junctions (Violin and Kholuyanov 1964, Slack 1965, Kroko and Milnes 1966, Blank 1969). Above this temperature care has to be taken to prevent sublimation of the bulk crystal, even in a SiC powder environment. Therefore, epitaxial growth (VPE and LPE) seems to be most adequate for the generation of p–n junctions in SiC.

The first investigations were centered on vapour-phase systems and many combinations of carbon- and silicon-rich compounds in varying ratios were tried. A typical system employs methyl-chlorsilanes or silicon tetrachloride and hydrocarbons (e.g. hexane) with hydrogen as carrier gas and a growth temperature of about 1850°C (Spielmann 1965, Bonnke and Fitzer 1966, von Münch and Pfaffeneder 1976). The growth rate is in the 0·05–0·15 $\mu m\,min^{-1}$ range. Nitrogen or ammonia gas doping yields n-type layers, while doping with diborane or trimethyl-aluminium is a convenient method of producing p-type layers. Because of the difficulties in the preparation of highly doped high-quality layers necessary for the production of SiC LEDs, LPE is preferred (Brander and Sutton 1969, Suzuki *et al* 1967a, von Münch *et al* 1976, Matsunami *et al* 1977).

The apparatus used for the deposition of SiC layers by LPE is shown in figure 2 and is explained in detail elsewhere (von Münch and Kürzinger 1978). The deposition of SiC layers takes place in a non-stoichiometric silicon–carbon melt under the influence of a

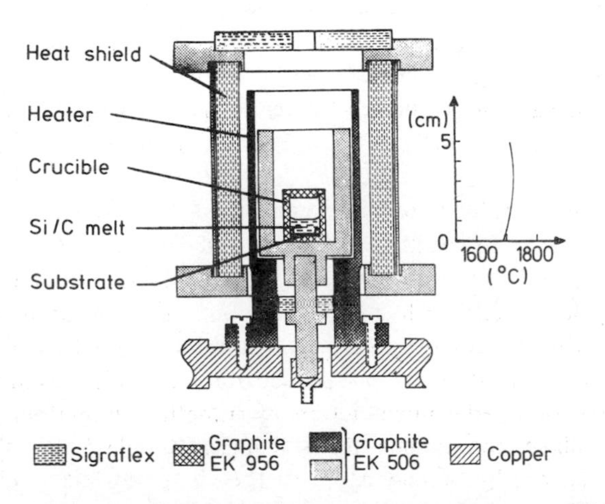

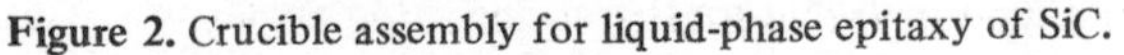

Figure 2. Crucible assembly for liquid-phase epitaxy of SiC.

temperature gradient. In the growth process some carbon is dissolved from the crucible wall, whereas SiC crystallisation is possible in the narrow cooler zone of supersaturated melt above the substrate. To maximise melt–vapour communication, the substrates are arrayed horizontally with a thin melt covering each one (Saul and Lorimor 1974). The most satisfactory growth and doping requires melts which are exposed to the ambient in a temperature range of 1600–1800 °C. This can be done by use of a half-closed crucible assembly which minimises the growth rate and yields a high-quality surface layer and uniform layer thickness.

The crucible assembly is positioned in such a manner that the temperature gradient at the substrate is about 10–30 $^{\circ}C\,cm^{-1}$, which enables crystal growth by diffusion transport and allows, with given growth temperature and cooling rate, a good control of the layer thickness with growth time. The temperature–time programme of a typical growth run in the preparation of overcompensated LEDs is shown in figure 3. A degassing and heat-up cycle (a) is followed by a growth period (b) at 1600 °C, with an argon pressure of 1 atm; the silicon melt is doped with aluminium. After this period the melt is cooled at a rate of about 15 $^{\circ}C\,min^{-1}$ for better control of layer thickness. During this period the system is evacuated, (c), and a nitrogen pressure of 5×10^{-4} Torr is established. Later on, the nitrogen pressure is raised to 5×10^{-2} Torr to form the p–n junction at position (d, e). Thus the SiC in the vicinity of the junction is compensated for, as could be proved by Hall and *C–V* measurements. An oxidised angle lap of a layer sequence prepared in this way showed a p-layer of about 5 μm followed by a compensated p-layer of nearly 1 μm and slightly compensated n-layer, also about 1 μm.

The same growth process can be used for the preparation of SiC MESFET layers with an intense degassing and evacuating cycle before establishing the growth temperature for the preparation of weak p^-(buffer) layers by adding a small amount of Al/Si. The active n-layer is deposited in the described cycle by establishing 5×10^{-4} Torr nitrogen during the cooling period. With a growth temperature of 1580 °C and the time programme described above it is possible to prepare an epilayer of about 0·5 μm and an

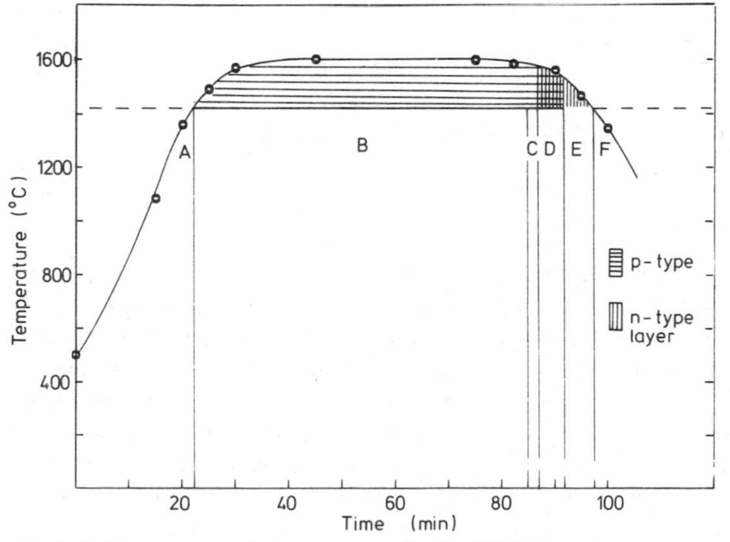

Figure 3. Temperature–time programme for the SiC LPE process.

electron concentration in the range of 5×10^{16}–$1 \times 10^{17}\,cm^{-3}$. The process yields a rather good p–n junction with a breakdown voltage exceeding 200 V and a quite smooth surface layer.

Moreover, it should be possible to prepare SiC Schottky–nn^+–IMPATT diodes in this manner as well as p^+nn^+–IMPATT diodes with a high p-type layer prepared by diborane doping. Whether it is possible to establish all the demands of IMPATT diodes (i.e. layer thickness and concentration of the breakdown region) has to be investigated.

In the case of microwave devices, such as SiC MESFETs with a very shallow n-type channel, it should be possible to develop an effective ion implantation process as is successfully done with GaAs devices. Besides the ability to establish the small channel depth over a great area, a fundamental advantage is the possibility of developing a full planar technology with selective ion implantation through a photoresist mask. This may be an important fact in the case of SiC considering mesa etching and gate recessing, which are rather difficult processing steps.

The ability to form n-type conducting layers in p-type hexagonal SiC by implanting ions from column V of the periodic table, especially nitrogen and phosphorus, was demonstrated in the early 1970s (Dunlap and Marsh 1969, Marsh and Dunlap 1970, Addamiano *et al* 1972). Although the authors had to use highly doped p-type substrates and high implantation doses (10^{13}–$10^{16}\,cm^{-2}$), annealing at 1400–1600 °C eliminated the radiation damage as was seen from the observed carrier mobility values; the surface carrier concentration of 10^{18}–$10^{21}\,cm^{-3}$ corresponded to $\frac{1}{2}$–$\frac{1}{3}$ of the implanted dose.

Further development in ion implantation should be possible by the use of high-resistivity substrates (as discussed previously), lower implantation doses to prepare n-layers with carrier concentrations about $10^{17}\,cm^{-3}$ and improved encapsulation during annealing; such experiments are now in progress.

4. Material qualifications

4.1. Microwave semiconductors

The fundamental question for the physical limits of future developments of semiconductor devices, such as packing densities, speed of operation, and power handling capability, results in some 'figures of merit' expressed by Johnson and Keyes:

$$Z_J = (4\pi)^{-1} E_b^2 v_{sat}^2 \qquad \text{(Johnson 1965)}$$

$$Z_K = \lambda (c v_{sat}/4\pi\epsilon)^{1/2} \qquad \text{(Keyes 1974).}$$

These figures of merit demonstrate that the power and frequency-limiting factors of transistors as well as the switching behaviour of semiconductor devices are dominated by the saturated drift velocity v_{sat}, the breakdown field E_b, the dielectric constant ϵ, and the thermal conductivity λ (c is the velocity of light). Some of these material parameters of SiC have been experimentally determined in the last two years and are summarised in comparison with some other microwave semiconductor materials in table 1.

The occurrence of a wide band gap is normally accompanied in semiconductors by a lower dielectric constant, an improved thermal conductivity and a higher breakdown field, and this is also true of SiC. It has been confirmed recently (van Opdorp and Vrakking 1969, von Münch and Pfaffeneder 1977) that SiC exhibits an extremely high

Table 1. Figures of merit for microwave semiconductors.

		Si	GaAs	SiC
μ_n	$cm^2 V^{-1} s^{-1}$	650	4000	250
μ_p	$cm^2 V^{-1} s^{-1}$	270	300	25
τ_n	ns			1 . . . 17
		0·1 . . . 10^6	0·1 . . . 100	
τ_p	ns			3 . . . 9
L_n	μm		6·2 . . . 15	
		10 . . . 10^3		0·4 . . . 1·9
L_p	μm		0·6 . . . 3	
v_{sat}	$cm\ s^{-1}$	1×10^7	2×10^7	2×10^7
E_b	$V\ cm^{-1}$	2×10^5	3×10^5	3×10^6
ϵ		12	11	9·7
λ	$W\ cm^{-1} {}^\circ C^{-1}$	1·5	0·5	3·5
Z_J	$V^2 s^{-2}$	$3{\cdot}2 \times 10^{23}$	$2{\cdot}9 \times 10^{24}$	$2{\cdot}9 \times 10^{26}$
Z_K	$W\ s^{-1} {}^\circ C^{-1}$	$6{\cdot}7 \times 10^7$	$3{\cdot}5 \times 10^7$	$2{\cdot}4 \times 10^8$
T_{max}	°C	250	420	>500

breakdown field, i.e. $3 \times 10^6\,V\,cm^{-1}$. In addition, there is evidence that the thermal conductivity normal to the C axis has a rather high room-temperature value of $3{\cdot}9\,W\,cm^{-1}\,{}^\circ C^{-1}$ and a value about 30% lower parallel to the C axis (Burgemeister *et al* 1979). The dielectric constant was determined to be 9·7 (Das and Ferry 1976). In addition to the relatively higher phonon energies of the wide band gap semiconductor, SiC has a saturated drift velocity of electrons higher by a factor of two than in silicon (von Münch and Pettenpaul 1977). The calculations of Johnson's and Keyes' figures of merit (table 1) demonstrate a considerable advantage of SiC in terms of power, frequency and switching behaviour compared with Si and GaAs and should stimulate the application of SiC for microwave devices with high-power handling capability such as MESFETS and IMPATT diodes. However, there may be restriction to higher series resistances (because of lower mobility) and higher parasitic capacitances if no semi-insulating substrates can be fabricated as is done in GaAs technology.

In addition, besides the lower mobility, a major drawback of SiC is the short lifetime of minority carriers (of the order of 10^{-9} s) and diffusion lengths of about 1 μm (Hoeck 1977), which restricted the fabrication of bipolar transistors (see table 1).

4.2. Light-emitting diodes

As an important figure of merit in the characterisation of light-emitting diodes, the external quantum efficiency η_{ext} provides the most basic information on radiative recombination processes in a LED. The external quantum efficiency consists of injection efficiency η_{inj}, light-producing efficiency η_{light} and optical efficiency η_{opt}:

$$\eta_{ext} = \eta_{inj} \times \eta_{light} \times \eta_{opt}.$$

An estimation of the injection efficiency of SiC light-emitting diodes is scarcely possible at present because cathodoluminescent spectra of the n- and p-regions of an overcompensation diode with best efficiency show that light generation takes place simultaneously in both regions, although the emission from the n-side is more efficient (Kürzinger *et al* 1978). Moreover, *I–V* characteristics with an ideality factor greater than 2 and rather high leakage currents at low voltages indicate considerable recombination in the space-charge region (which could be expected with short diffusion lengths) and at surface defects.

The probability of radiative recombination of electrons and holes depends in the first place on the band structure of the semiconductor. SiC has a silicon-like band structure, but with a large unit cell (12 atoms) and hence small Brillouin zones with a high density of bands (Moss *et al* 1973). The absorption of 6H SiC was measured for photon energies up to 4·9 eV; indirect absorption edges were found at 3·0, 3·7 and 4·1 eV, respectively (Choyke and Patrick 1968). These facts do not indicate efficient near band gap luminescence by recombination at donor or acceptor levels as in the case of GaAs and associated ternaries like GaAsP and GaAlAs, nor the efficient recombination by isoelectronic traps as in the case of the green luminescence of GaP : N diodes, for example. The reason for the latter is the large intraband distance between the Γ- and X-point of the valence band (Choyke and Patrick 1968). At other than very high temperatures the phonon term should be most significant and results in a reduction of radiative recombination for this indirect-gap material.

The light-emitting efficiency is determined by the number of radiative recombination centres and by the effectiveness of non-radiative centres in terms of radiative τ_r and non-radiative τ_{nr} recombination lifetimes, and can be expressed after simplification as:

$$\eta_{light} = \frac{1/\tau_r}{1/\tau_r + 1/\tau_{nr}}.$$

As a result, the excess minority carrier lifetime of radiative processes should be short and that for non-radiative processes long. Contrary to this a very short electron and hole lifetime in the range of 1–17 ns for non-radiative processes was measured, whereas overcompensated LEDs showed a somewhat higher value which can be attributed to a radiative trap centre with an activation energy of about 320 meV (Hoeck 1977), probably associated with the aluminium acceptor.

It can be supposed that non-radiative recombination processes in connection with deep impurities and crystalline defects, particularly silicon vacancies, cause a low radiative efficiency in the indirect-gap semiconductor SiC. A different number of silicon vacancies could be an explanation for the higher efficiencies of LPE-prepared LEDs compared with those prepared by VPE, as was observed with GaP : N diodes. Furthermore, in highly doped SiC LEDs a non-radiative three-particle process is possible with a relatively short lifetime of the order of nanoseconds, thus reducing the efficiency of the indirect radiative recombination by a factor of about 10^{-3} (Brander 1972).

Moreover, thermal conductivity is important, since the efficiency of luminescence falls rapidly with increasing temperature. Compared with highly doped SiC (Burgemeister *et al* 1979) the other optoelectronic materials have a rather low thermal conductivity (table 2), which is a restriction to low current levels.

Finally, care will have to be taken in order to achieve an external efficiency comparable with the efficiency with which light is produced internally. The main loss mechanism is of course absorption, either directly in the material and contacts, or after reflection at

Table 2. Figures of merit for optoelectronic materials.

		GaAs	GaP	SiC
Band gap type		direct	indirect	indirect
E_g	eV	1·4	2·3	2·9
E_{phon} (LO)	meV	36	50	120
α Band–band		>1000	>100	20–40
α Shallow impurity	cm^{-1}	~1000	~100	
α Deep impurity		50	<5	
λ	$W\ cm^{-1}\ {}^\circ C^{-1}$	0·44	0·77	2·5
n		3·6	3·3	2·7

the crystal–air interface, i.e. the absorption coefficients, α, and refractive indices, n, of semiconductor materials are significant features of the optical efficiency. Unfortunately the commercial optoelectronic materials have relatively high refractive indices (table 2) which result in a small angle of total reflection. Wide band gap semiconductors show advantages in this respect because of their smaller refractive indices, for instance $n = 2{\cdot}7$ in the case of SiC (Thibauld 1944).

A further advantage in comparison with direct-gap materials is a lower absorption coefficient, especially if the recombination is caused by deep impurities with an active layer on a transparent substrate, as in the case of GaP:Zn,O diodes (table 2). In SiC the absorption coefficient is about 20–40 cm^{-1} (Brander 1972, Kürzinger 1977) for crystals with carrier concentrations of $10^{17} cm^{-3}$. In the blue SiC LEDs described below, having n-type layers with carrier concentrations of $5 \times 10^{19} cm^{-3}$ in the light output layer, the absorption coefficient may have a somewhat higher value.

4.3. Further developments

In the future 'microprocessor-controlled technical world' an old and new device will become important, namely the sensor. In view of this, SiC is a potentially useful material because of its extreme chemical inertness, the possibility of operation up to 500–600 °C for n-type and considerably higher temperatures for p-type material, its high radiation damage resistance and high power density capability (Campbell and Berman 1969, Mehrwald 1968). These properties open the possibility of using unpackaged devices in aggressive environments (i.e. in the chemical industry, nuclear industry, and astronautics), high-temperature applications and power current technology.

5. Applications

In the following section two attractive applications will be discussed in detail, i.e. blue light-emitting diodes and Schottky-barrier field-effect transistors.

5.1. Device technology

The process steps for the fabrication of both devices in mesa technology are similar. An oxidation step in wet oxygen at 1070°C yields an oxide layer of about 4000 Å on the $(000\bar{1})$ basal plane (carbon face) of the epitaxial layer. The mesa regions are defined by conventional photolithographic techniques. As a result of its enormous hardness, the shaping of SiC has to be done by selective (SiO_2-masked) gas etching in a mixture of 72 vol.% Ar, 22 vol.% Cl_2 and 6 vol.% O_2 at 1050 °C (Campbell and Berman 1969); the etching rate is about 3 $\mu m\ min^{-1}$. This demonstrates a major drawback compared with GaAs MESFETS where gate recessing by wet etching through a photoresist mask is possible before the metallisation process.

The ohmic contacts for n-type material consist of a thin (1000 Å) nickel layer and are made by a (SiO_2-supported) lift-off technique and sintered at 1000 °C in vacuum. Contacts to the p-type substrate are made by alloying with Al/Si eutectic at 950 °C. In either case a strengthening of the contact pattern with Cr/Au or Ti/Pt/Au metallisation is necessary to achieve easy bonding. The Schottky gate contact of the MESFET is generated by depositing Ti/Pt/Au (4000 Å) with a lift-off technique, as in the case of GaAs MESFETS where it is used as an alternative to aluminium.

The wafers contain approximately 400 diodes of 0·25 mm^2 squares and mesas of 400 and 300 μm in diameter, whereas the MESFET chip dimensions are 0·35 × 0·35 mm. After dicing with a Tempress saw each chip is mounted with conductive epoxy resin on a TO-18 package with a suitable reflector in the case of the LEDs and in a small metal/ceramic microwave package in the case of the MESFETS. The LED chips are covered with a clear epoxy lens after the wire-bonding operation, which gives an external efficiency increase of two to three times.

5.2. Blue-emitting diodes

VPE and LPE diodes (n-type layers on p-type substrates) have been produced by Brander and Sutton (1969), Suzuki *et al* (1976a), von Munch *et al* (1976), Matsunami *et al* (1977),

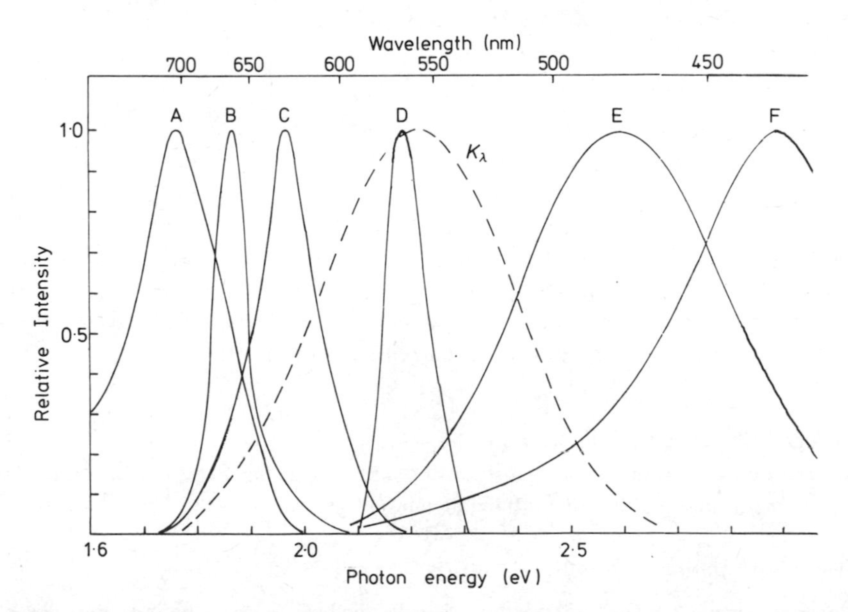

Figure 4. Emission spectra of various light-emitting diodes: A, GaP : Zn, O; B, $GaAs_{0\cdot6}P_{0\cdot4}$; C, $GaAs_{0\cdot15}P_{0\cdot85}$: N; D, GaP : N; E, SiC : Al, N overcompensated LED; and F, SiC : Al, N two-step LED.

von Münch and Kürzinger (1978) and Kürzinger *et al* (1978). Figure 4 shows the emission spectra of two of our SiC diodes (curves E and F) compared to some commercially available GaP and GaAsP LEDs (Plihal and Weyrich 1976). The emission spectrum according to curve E results from diodes which were produced by the overcompensation method described previously, whereas curve F results from diodes which were produced by a two-step LPE process whose junction region is nearly free from compensation. This spectrum is markedly shifted to higher energies.

Room-temperature emission spectra of overcompensated diodes with different hole concentrations in the luminescent layer show that the highest external quantum efficiency is obtained from p-type layers with an aluminium concentration corresponding to a hole density around $5 \times 10^{18}\,\mathrm{cm}^{-3}$; the efficiencies of these diodes are nearly an order of magnitude higher than those of the two-step diodes (von Münch and Kürzinger 1978). Together with the better overlap of the emission spectra of overcompensated diodes with the human eye sensitivity curve an appreciably higher luminous efficiency could be reached.

Both types of diode show a very broad band (about 75 nm) at room temperature which results usually from the superposition of Al–N pair luminescence (emission peak at 2·6 eV), transitions via aluminium acceptors (emission peak about 2·75 eV) and phonon-assisted band–band recombination (emission peak at 2·9 eV), and a low degree of colour saturation of approximately 50% compared with nearly 100% in the GaP and GaAsP diodes.

The peak energy of the dominant transition depends on carrier concentration (i.e. also the compensation of the p- and n-type layers), crystal temperature and injection current. In the case of the high-efficiency overcompensated diodes, it is assumed that the dominant radiative recombination mechanism at room temperature is a recombination at distant N donor–Al acceptor pairs which would be consistent with the activation energies obtained by Hall measurements.

A marked shift towards the high-energy end of the spectrum is observed when the diodes are heated (von Münch and Kürzinger 1978, Kürzinger *et al* 1978) or at very high injection levels (Brander and Sutton 1969). From the photon energy values it must be concluded that band–band transitions start to dominate, as in other optoelectronic materials. The external quantum efficiency only slightly decreases with increasing temperature. However, due to the shift of the spectrum to shorter wavelengths, the brightness of the diodes decreases much faster.

The discussion of §4 implies that in the interest of achieving a high external efficiency, low absorption can only be reached with a thin layer of material between the junction and the diode surface. About 10 diodes from one wafer were selected, with the criterion of similar I–V characteristics, for the examination of the relation between the thickness of the light-output layer and the external efficiency. The layer thickness of each diode was determined by measuring an oxidised angle lap of the respective neighbour element. Figure 5 demonstrates that the optimum thickness of the light-output layer should be about $1\,\mu\mathrm{m}$, which corresponds to the short diffusion length. In the case of the very thin n^+ layer of $0{\cdot}4\,\mu\mathrm{m}$, the current was restricted to the area of contact and the majority of the light could not escape.

A similar examination shows that the active p-layer should not exceed $10\,\mu\mathrm{m}$, because diodes with a layer thickness of 20–30 $\mu\mathrm{m}$ yield lower efficiency as a result of poor layer

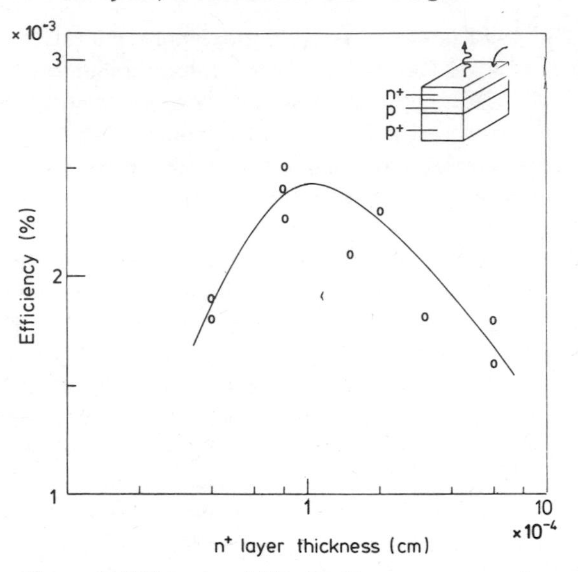

Figure 5. Efficiency of SiC:Al, N overcompensated LEDs with different thickness of the light-output layer (active p-layer: 4 μm; diode current: 50 mA; Acheson substrate).

and surface quality. Concerning the latter, the active p-layer should not be too thin since it is important to avoid contamination of the active layer by impurities from the substrates. Although there are some indications of the advantage of Lely substrates compared with Acheson substrates, it cannot be quantified. Moreover, the use of low-doped transparent SiC substrates could improve the optical coupling efficiency, as is spectacularly demonstrated in the case of orange–red GaAsP diodes.

Another important task is the reduction of harmful non-radiative centres coupled with improvements in the growth process, which should also permit a better control of the doping concentration in the vicinity of the p-n junction in order to improve the injection efficiency. The generation of radiation of energy greater than the applied bias (2·8 V) is attributed to degenerate material, where tunnelling of carriers across the barrier, assisted by a non-radiative Auger process, becomes important at low biases. Concerning this fact, capacitance measurements demonstrate that profiles with shallow gradients of doping and low effective carrier concentration yield high efficiency. These last results need a more detailed examination and interpretation.

The light flux emitted by an average-quality SiC LED compared to those of the commercially available GaP and GaAsP LEDs (Plihal and Weyrich 1976) is shown in figure 6. At the present state of development a two orders of magnitude lower luminous efficiency of SiC diodes is observed at the same diode current; their light output can be improved by operating at a considerably higher current level. The highest external quantum efficiency observed so far at room temperature is 6×10^{-5}, i.e. three times more efficient than the SiC diode indicated in figure 6 (Kürzinger *et al* 1978).

5.3. Field-effect transistors

SiC Schottky-barrier FETs with 10 μm gate-length have been produced recently (von Münch *et al* 1978b). For this case devices with an active gate-length of 1·1 ± 0·1 μm, a gate-width of 300 μm and a source–drain distance of 3·5 μm have been fabricated. From an oxidised angle lap measurement, it can be stated that the epitaxial layer thickness is of the order of 0·5 μm. The sheet resistance of the channel is about 500 Ω/□.

DC measurement results of these SiC devices compared with similar GaAs FETs with the same geometrical structure, sheet resistance and saturation current are summarised in table 3. The maximum transconductance of SiC FETs is found to be 3·0 mA V^{-1}, i.e. a factor of 8 lower than in GaAs. The reason is to be seen in the considerably higher source–drain resistance of SiC FETs (about 500 Ω, including a contact resistance of 102 Ω) as a result of the lower mobility values. Both values can be reduced to GaAs level if highly doped layers are used for the ohmic regions of the device. The pinch-off voltage amounts to 10 V, although a complete pinch-off with a current of 10 μA was observed. Sometimes an incomplete current saturation was observed, with differences from wafer to wafer.

In general there are two methods to realise a power FET with 1 W output or more. The first is to design a FET pattern with a 'wide' gate which can control a large drain current. Another way of increasing the output power is to maximise the drain–source breakdown voltage in order to apply higher bias voltages. Because of their restricted breakdown field, GaAs FETs are composed of many small-signal FETs connected in parallel with a gate-width of 5000 μm or more, corresponding to a saturation current exceeding 1 A and appropriate transconductance with drain–source breakdown voltages of nearly 20 V (Suzuki *et al* 1976b). The drawbacks of these methods are technological inadequacies in terms of non-uniformity of the GaAs epitaxial layer and RF power distribution among component MESFETs. In the case of SiC power FETs these problems can be solved because of the considerably higher drain–source voltage which restricts the gate-width to lower values. Drain–source breakdown voltages of 80 V have been reached so far and should be improved by proper device technology. The high-frequency of SiC FETs evaluated by means of scattering-parameter and noise figure measurements is being examined and the results will be published in the future.

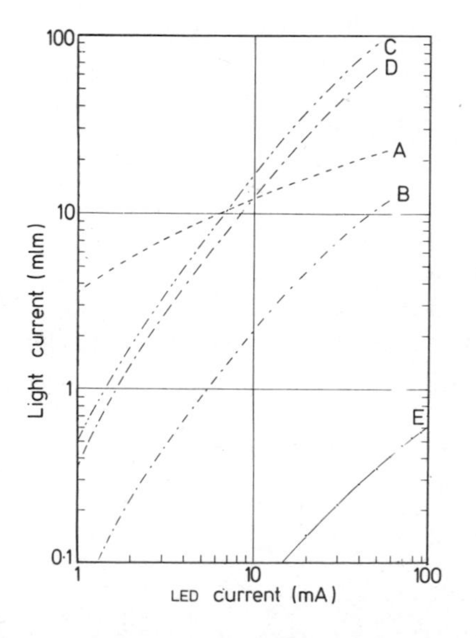

Figure 6. Light output of various light-emitting diodes against current: A, GaP:Zn, O; B, $GaAs_{0\cdot6}P_{0\cdot4}$; C, $GaAs_{0\cdot15}P_{0\cdot85}$:N; D, GaP:N; and E, SiC:Al, N overcompensated LED.

Table 3. Electrical characteristics of GaAs and SiC MESFETs (DC performance).

Symbol	Parameters and test conditions		Units	GaAs	SiC
ρ/d	Sheet resistance		$\Omega/\square$	461	536
I_{DSS}	Saturated drain current	$V_{DS} = 5, 10$ V $V_{GS} = 0$ V	mA	12	12
V_P	Pinch-off voltage	$V_{DS} = 5, 10$ V $I_{DS} < 100\ \mu$A	V	2·0	10
V_{sat}	Saturation voltage		V	0·8	8·0
R_{SD}	Source–drain resistance		Ω	20	500
U_{DS}	Source–drain voltage	$I_{DS} < 100\ \mu$A	V	12	>80
U_{GD}	Gate–drain voltage	$I_{DG} < 100\ \mu$A	V	8	80
g_m	Transconductance	$V_{DS} = 5, 10$ V $V_G = 0$	mA V^{-1}	24	3·0

6. Conclusion

At the present state of development the most important semiconductor parameters of SiC have been determined and demonstrate the high potential for special applications such as blue light-emitting diodes, microwave power devices and sensors in aggressive environments. A mesa technology was determined which permits the fabrication of devices in laboratory scale. Nevertheless, the future of SiC as a semiconductor depends strongly on the progress of single crystal growth, which is the key to commercial success.

References

Acheson A G 1892 *British Patent No.* 17911

Addamiano A, Anderson G W, Comas J, Hughes H L and Lucke W 1972 *J. Electrochem. Soc.* **119** 1355

Berman I and Ryan C E 1971 *J. Cryst. Growth* **9** 314

Blank J M 1969 *Silicon Carbide 1968* (New York: Pergamon) p179

Bonnke N and Fitzer E 1966 *Ber. Dt. Keram. Ges.* **43** 180

Brander R W 1972 *Phys. Rev. Tech.* **3** 145

Brander R W and Sutton R P 1969 *J. Phys. D: Appl. Phys.* **2** 309

Burgemeister E A, von Münch W and Pettenpaul E 1979 *J. Appl. Phys.* **50** *5790*

Campbell R B and Berman H S 1969 *Silicon Carbide 1968* (New York: Pergamon) p211

Choyke W J and Patrick L 1968 *Phys. Rev.* **172** 769

Das P and Ferry D K 1976 *Solid-St. Electron.* **19** 851

Dunlap H L and Marsh O J 1969 *Appl. Phys. Lett.* **15** 311

Griffiths L B and Mlavski A I 1964 *J. Electrochem. Soc.* **111** 805

Halden F A 1960 *Silicon Carbide 1960* (New York: Pergamon) p115

Hoeck P 1977 *Dissertation* Technische Universität, Hannover

Johnson A 1965 *RCA Rev.* **26** 163

Kapteyns C J and Knippenberg W F 1970 *J. Cryst. Growth* 7 20

Kendall J T 1953 *J. Chem. Phys.* **21** 821

Keyes R W 1974 *Silicon Carbide 1973* (Columbia, SC: University of South Carolina Press) p534

Knippenberg W F 1963 *Philips Res. Rep.* **18** 205

Knippenberg W F and Verspui G 1966 *Philips Res. Rep.* **21** 113

Kobayashi F, Ikawa K and Iwamoto K 1975 *J. Cryst. Growth* **28** 395

Kroko L J and Milnes A G 1966 *Solid-St. Electron.* **9** 1125

Kürzinger W 1977 *Dissertation* Technische Universität, Hannover

Kürzinger W, von Münch W, Pettenpaul E and Theis D 1978 *IEEE LED Spec. Conf. San Francisco*
Lely J A 1955 *Ber. Dt. Keram. Ges.* **32** 229
Marsh O J and Dunlap H L 1970 *Radiat. Effects* **6** 301
Matsunami H, Ikeda M, Suzuki A and Tanaka T 1977 *IEEE Trans. Electron. Devices* **24** 958
Mehrwald K H 1968 *Ber. Dt. Keram. Ges.* **45** 76
Moss T S, Burell G J and Ellis B 1973 *Semiconductor Opto-Electronics* (London: Butterworth)
von Münch W, Hoeck P and Pettenpaul E 1978b *Int. Electron Devices Mtg, Washington, DC, 1977* (New York: IEEE) p337
von Münch W and Kürzinger W 1978 *Solid-St. Electron.* **21** 1129
von Münch W, Kürzinger W and Pfaffeneder I 1976 *Solid-St. Electron.* **19** 871
von Münch W and Pettenpaul E 1977 *J Appl. Phys.* **48** 4823
——1978 *J. Electrochem. Soc.* **125** 294
von Münch W and Pfaffeneder I 1976 *Thin Solid Films* **31** 39
——1977 *J. Appl. Phys.* **48** 4831
van Opdorp C and Vrakking J 1969 *J. Appl. Phys.* **40** 2320
Plihal M and Weyrich C 1976 *Siemens Z.* **50** 167
Potter R M and Sattele J H 1972 *J. Cryst. Growth* **12** 245
Round H J 1907 *Electrical World* **19** 309
Saul R H and Lorimor O G 1974 *J. Cryst. Growth* **27** 183
Scace R I and Slack G A 1959 *J. Chem. Phys.* **30** 335
Slack G A 1965 *J. Chem. Phys.* **42** 805
Spielmann W 1965 *Z. Angew. Phys.* **19** 93
Suzuki A, Ikeda M, Nagano N, Matsunami H and Tanaka T 1976a *J. Appl. Phys.* **47** 4546
Suzuki H, Suyama K and Fukuta M 1976b *Fujitsu Sci. Tech. J.* **12** 163
Thibauld N W 1944 *Am. Miner.* **29** 327
Violin E E and Kholuyanov G F 1964 *Sov. Phys.–Solid St.* **6** 465

Ion beam, plasma and reactive ion etching

C J Mogab
Bell Laboratories, Murray Hill, New Jersey 07974, USA

Abstract. The application of ion beam, plasma and reactive ion etching for pattern transfer in the fabrication of electron devices is reviewed with emphasis on recent developments. The need for highly anisotropic etching with adequate selectivity to complement high resolution lithography is pointed out. A brief description of typical equipment and operating conditions is given and some of the physical and chemical effects bearing on resolution and selectivity are discussed for each of these methods. Finally, some consideration is given to applications, advantages and limitations of the techniques.

1. Introduction

New methods for fabricating microelectronic devices have evolved rapidly over the past several years in response to the demand for larger scale integration to effect cost savings and enhance performance. Improvements in lithographic techniques have enabled replication of high density patterns of micron and even sub-micron size features in photo or electron sensitive resist masks. The transfer of such patterns into underlying inorganic materials, by etching, requires a high degree of etch anisotropy in order to retain pattern fidelity. Liquid phase chemical etching, which is usually isotropic, has thus given way to the use of gas phase, plasma-assisted etching methods which, under certain conditions, can produce very high resolution pattern transfer. The anisotropy results either directly or indirectly because charged particles incident on the surface arrive from the plasma directed by an imposed electric field and this can produce a vertical etch rate which substantially exceeds the lateral etch rate.

Plasma-assisted etching methods include ion etching (ion milling and sputter etching), reactive ion etching (reactive sputter etching) and plasma etching. These techniques have in common the use of plasma activated species to achieve etching. They differ, however, in the mechanisms involved in the etching process. Ion etching relies on *physical sputtering* by energetic noble gas ions (several hundred to several thousand eV). Plasma etching can be *defined* as the formation of a volatile product upon *chemical reaction* between a solid and active radicals created in a molecular gas discharge. Reactive ion etching (RIE) lies somewhere between these extremes, that is, both physical and chemical effects contribute to etching. In practice, the distinction between plasma etching and RIE is usually drawn on an operational rather than a mechanistic basis. Specifically, plasma etching is taken to mean formation of a volatile product in a molecular gas discharge at relatively *high pressure* ($\sim 10^{-1}$–10 Torr) with the substrate either grounded or electrically floating. RIE implies formation of a volatile product in a molecular gas discharge at relatively *low pressure* ($\sim 10^{-3}$–10^{-1} Torr) and with the substrate driven at the operating potential. In

0305-2346/80/0053-0037 $03.00

some cases there may be little or no difference in *actual substrate environment* for these methods and comparable results may be obtained. Accordingly, both physical and chemical effects may contribute to etching under the operating conditions used in plasma etching as given above. In such instances the difference between plasma etching and reactive ion etching may be in the degree to which the various chemical and physical effects contribute rather than a distinct mechanistic one. Thus, although plasma etching may be defined as a purely chemical process the term is often used to describe the results of cooperative physical and chemical processes.

In this paper we will review, briefly, the motivation for developing high resolution etching, the types of equipment used in plasma-assisted etching, some of the pertinent phenomena observed and the advantages and limitations of the various techniques. A detailed account of plasma-assisted etching published recently (Melliar-Smith and Mogab 1978) includes extensive references to the literature on the subject through to the early part of 1978. Consequently, the present paper will concentrate mainly on developments since that time.

2. Requirements for high resolution etching of microelectronic devices

The term 'resolution' as applied to pattern transfer implies a measure of fidelity which can be quantified in terms of etch bias and tolerance. Etch bias is the difference in lateral dimension between an etched image and the original mask image. Tolerance is a measure of the distribution of bias values and is an indicator of feature size control. Figure 1 illustrates two extremes of edge profile: isotropic etching results in a bias which is equal to twice the etched film thickness (assuming no overetching) while fully anisotropic etching has zero bias. The difference between these cases is insignificant provided the ratio of feature size (lateral dimension) to film thickness is large. When this ratio drops below about 5 : 1, however, the undercutting characteristic of isotropic etching becomes, at the least, undesirable and, at the extreme, intolerable. As an illustration of the problem consider etching an array of equal lines and spaces (or an isolated window) of dimension x in a film of thickness t. To compensate in the masking layer for isotropic undercutting, the space between mask features (or the window diameter) must be $x - 2t$. Clearly, if x is of the same order as $2t$ this places unrealistic demands on the resolution of the lithographic mask generation technique. It would be impossible, for example, to compensate a contact window mask to produce 2 μm diameter windows in a 1 μm thick intermediate insulator for a Si integrated circuit if isotropic etching was to be used. In general then,

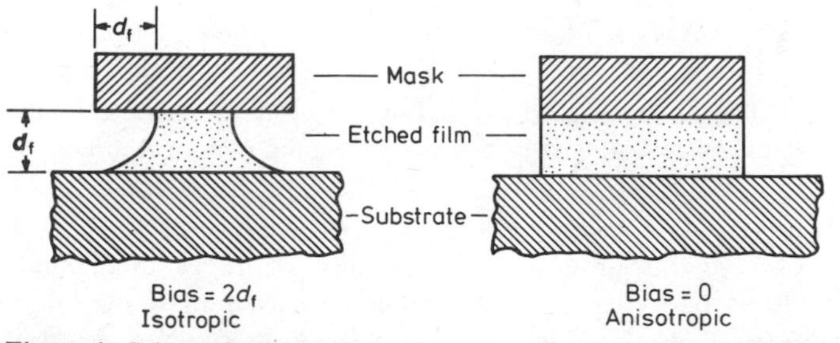

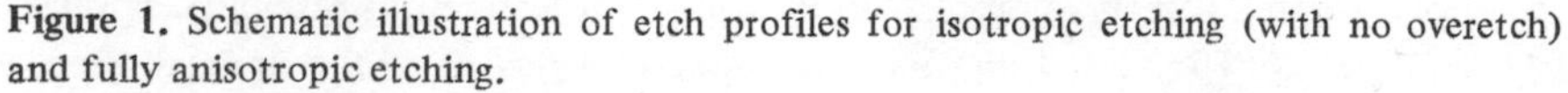

Figure 1. Schematic illustration of etch profiles for isotropic etching (with no overetch) and fully anisotropic etching.

as the ratio of feature size to film thickness decreases the degree of etch anisotropy required increases. Present and future generations of semiconductor integrated circuits, magnetic bubble memories and integrated optical devices will require high resolution means for pattern transfer (Ephrath and Mogab 1979); without question this has been the primary motivation for the heavy emphasis on the development of plasma patterning methods.

Etch selectivity is an equally important aspect of pattern transfer. Ideally, both the etch mask and underlying material exposed by etching should be unetchable. Any lateral etching of the mask will result in loss of linewidth. The extent to which attack of underlying material is permissible depends on the specific device being fabricated. It is a simple matter to calculate the minimum selectivity (defined as etch rate ratio) required, given the mean film thickness, $\bar{d}_f$, and the maximum tolerable penetration into the underlying material, d_{sm}. Results of such calculations are shown in figure 2 for a typical case with film thickness uniformity of 5% and an etch rate uniformity of 10%. An important feature of figure 2, in the context of anisotropic etching, is the strong dependence of the required selectivity on the degree of overetch. It often occurs in device fabrication that films to be etched cover stepped topography and thus have 'extra' vertical thickness as illustrated in figure 3. For highly anisotropic etching, this means that when the film on planar surfaces has been completely removed (thereby exposing underlying material) etching must be continued to remove the additional thickness labelled 'residue' in figure 3. In effect, overetching is required and is the price exacted for high resolution. (Note that

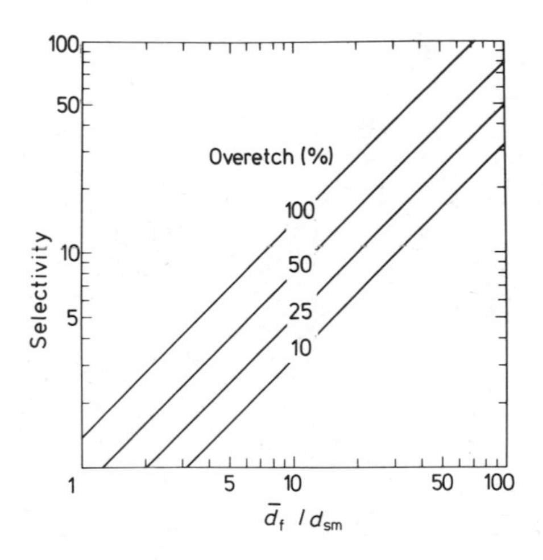

Figure 2. Plot showing the minimum etch selectivity (defined as etch rate ratio of film to substrate) as a function of the ratio of mean film thickness, $\bar{d}_f$, to maximum permissible penetration into the underlying substrate, d_{sm}. 5% film thickness uniformity, 10% etch rate uniformity.

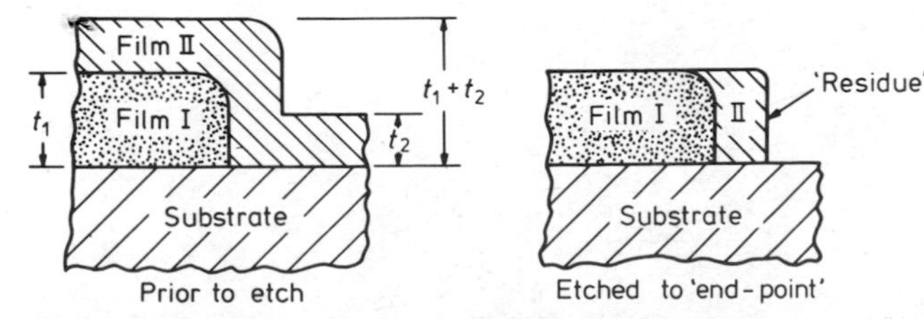

Figure 3. Schematic illustration of residual film at the 'end' of anisotropic etching due to the 'extra' thickness caused by stepped topography.

the extra thickness would not result in any 'residue' if etching was isotropic.) Perfectly uniform films and perfectly uniform etching would shift the curves in figure 2 downward by a factor of 1·47, but overetching would still be necessary with stepped surfaces. A 100% overetch would nominally be required for fully anisotropic etching of a film passing over a step of height equal to the film thickness. This is a quite commonly encountered situation in MOS integrated circuit fabrication; for example, in the etching of Al metallisation or the second level of a two-level polycrystalline Si gate structure. Typically, the ratio $\bar{d}_f/d_{sm}$ is about 10 in these cases and one can see from figure 2 that a selectivity of about 16 is required, whereas for the same conditions without steps a selectivity of 3 would suffice.

3. Equipment for plasma-assisted etching

The low pressure gas discharges used in ion, plasma and reactive ion etching can be generated in a variety of ways, however, certain basic designs seem to be favoured. Detailed descriptions of apparatus can be found in Melliar-Smith and Mogab (1978) and Melliar-Smith (1976). Here we will present only a brief discussion.

Ion milling systems consist of an ion source physically separated from the substrates by a set of grids biased so as to extract an ion beam from the source which then impinges on the substrates. The beam can be neutralised by a thermionic filament as required. A typical system is illustrated schematically in figure 4. Commercial systems are available with uniform beams up to 25 cm in diameter. When an inert gas is employed in the ion source, etching occurs by physical sputtering.

Sputter etching is usually done in an unconfined RF discharge illustrated schematically in figure 5. The substrates are placed on the driven electrode (cathode) whose area is much smaller than the grounded surface area of the system to ensure that most of the voltage drop occurs across the ion sheath at the cathode (Koenig and Maissel 1970, Coburn and Kay 1972). This sheath is parallel to the cathode surface. Ions present in the body of the plasma are accelerated normal to the sheath and transfer their momentum to the surface on impact. When sufficient voltage is applied, sputtering

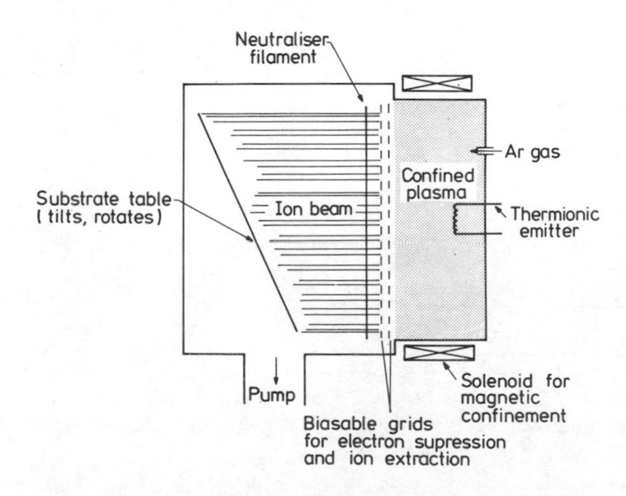

Figure 4. Schematic illustration of an ion beam etching system.

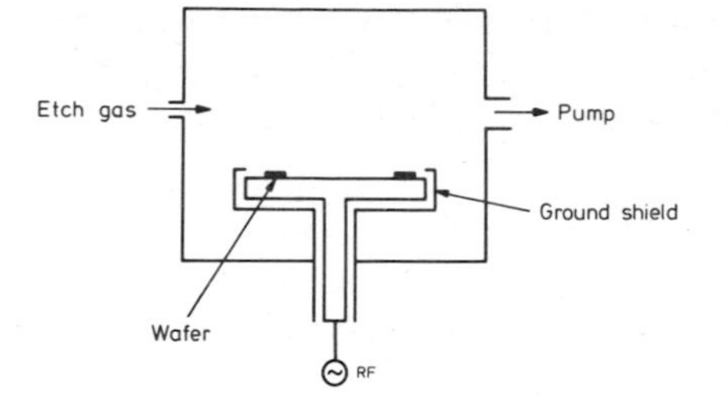

Figure 5. Schematic illustration of RF diode system employed in sputter etching and RIE.

results. If noble gases are used, the etching process is a purely physical one. Such systems are readily converted to a RIE mode by replacing the inert gas with a molecular gas containing constituents which can form volatile products with the substrate material and perhaps by changing the materials of construction somewhat. The latter change is made to avoid sputter deposition of involatile material on the substrates. This effect will be discussed further on. The molecular gases chosen invariably contain one or more halogen atoms since halides are numerous among the simple inorganic compounds which are volatile at temperatures low enough to be useful for pattern transfer. Halocarbons, in particular, have been used extensively (Melliar-Smith and Mogab 1978).

Similar gases are employed for plasma etching, but the reactor configuration is different. Two types of reactors are prevalent. So-called barrel reactors (see Melliar-Smith and Mogab 1978 for an illustration) cannot generally be used for highly anisotropic etching (Melliar-Smith and Mogab 1978) and thus are not normally employed for fine line patterns. These systems provide a good example of plasma etching as defined earlier. That is, the etching occurs essentially entirely by chemical reaction with neutral fragments from the discharge. For example, when a mixture of CF_4 and O_2 is used, copious production of F atoms occurs in the plasma (Melliar-Smith and Mogab 1978). These atoms are relatively long lived and react spontaneously with Si and its compounds to form volatile SiF_4, even when the substrates are remote from the plasma. Not surprisingly, the etching in such systems is largely isotropic.

The second type of apparatus uses simple planar, parallel plate electrodes as illustrated schematically in figure 6. The system is similar to the RIE version except that: (1) the electrodes are more nearly symmetric resulting in greater confinement of the plasma; (2) the substrates are placed on the grounded electrode and, as mentioned before; (3) the operating pressure is generally higher. The high degree of confinement of the plasma in

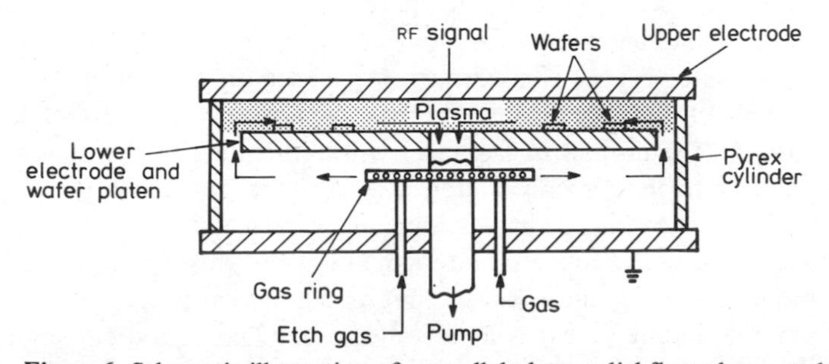

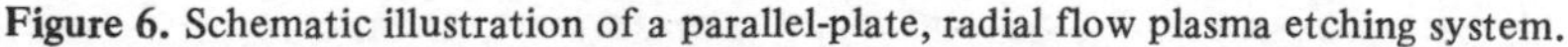
Figure 6. Schematic illustration of a parallel-plate, radial flow plasma etching system.

these systems can result in high plasma potentials (Coburn and Kay 1972, Vossen and Cuomo 1978, Vossen 1979). From the viewpoint of the energy of charged particles incident on the surfaces of substrates, it is the difference between the plasma and surface potentials that counts, irrespective of the location of the substrates (that is, whether on driven or grounded electrode). Available evidence indicates that this difference can be as large as several hundred volts in parallel plate, plasma etching systems (Vossen 1978). Accordingly, the effects of energetic ion or electron bombardment must be considered in these systems as well as in RIE systems.

Owing to the interest in high resolution etching, all further discussion of plasma etching will be limited to parallel plate-type systems.

4. Some phenomena pertinent to plasma-assisted etching

4.1. Ion etching

Rare gas ion etching is by far the best understood of the plasma patterning methods. Regardless of system configuration, the same basic phenomenon, namely sputtering, is solely responsible for etching. Sputtering occurs when an incident energetic atom or ion transfers its momentum to one or more substrate atoms such that the momentum of the substrate atoms is vectored away from the surface and the energy imparted to them exceeds their binding energy.

It is apparent that the plasma, in this case, serves merely as a source of ions. The ion energy can be controlled directly, for ion milling, by control of grid potentials and, somewhat less directly, in sputter etching, by the power coupled into the plasma. (We assume here that RF excitation of the plasma is used in the latter case. See Melliar-Smith and Mogab (1978) for the advantages of doing this.) Despite the common mechanism there are important differences between these techniques. In particular, the ions impact the substrates at normal incidence in sputter etching whereas the angle of incidence can be varied at will with ion milling. Also, the operating pressure at the substrates is usually an order of magnitude lower with ion milling and the beam current and energy are subject to direct control.

Etch rates are dependent on the flux and energy of particles arriving at the surface and their angle of incidence. The latter dependence results from the fact that momentum is transferred. Ions arriving at oblique incidence have to undergo less directional change of momentum to eject atoms in a forward direction. Typical data illustrating the variation of etch rate with angle of incidence are shown in figure 7 for several materials employed in the fabrication of magnetic bubble memories.

Notice that the etch rate reaches a maximum for angles off normal incidence. This leads to an important etching phenomenon known as *faceting.* This effect is shown schematically in figure 8. The normal to the facet will be inclined to the incoming ions at an angle corresponding to the maximum etch rate in figure 7. Clearly, if the facet in the mask propagates into the film during etching, resolution will be compromised. A relatively thick mask or one with a low etch rate would prevent this. Ensuring a low mask etch rate is not simple, however, since ion etching is a physical process and the selectivity, except in certain unique cases, is relatively poor. (This is sometimes done by mixing O_2 with the sputtering gas and using a mask made from a readily oxidised metal

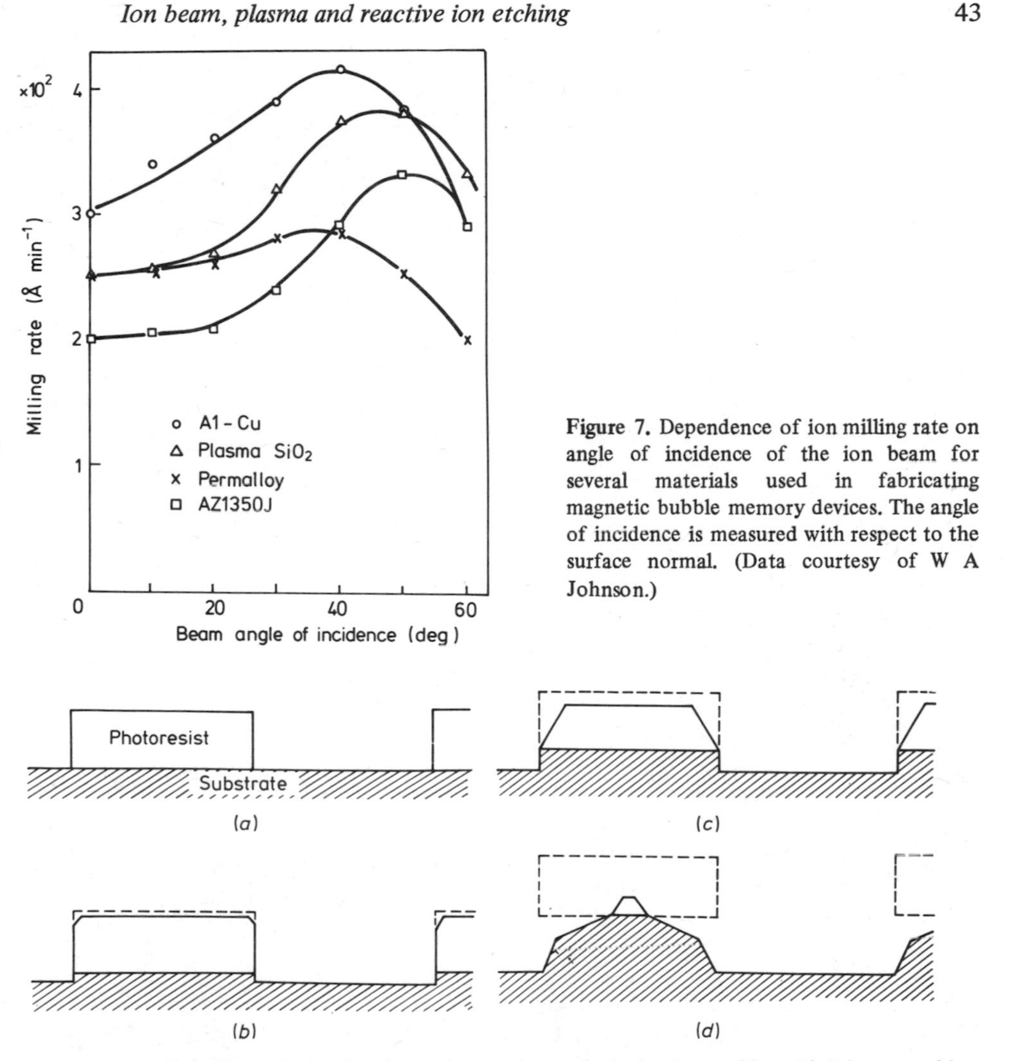

Figure 7. Dependence of ion milling rate on angle of incidence of the ion beam for several materials used in fabricating magnetic bubble memory devices. The angle of incidence is measured with respect to the surface normal. (Data courtesy of W A Johnson.)

Figure 8. Facet formation in a photoresist mask during ion etching. (*a*) Prior to etching; (*b*) initiation of the facet; (*c*) facet intersects substrate surface; (*d*) substrate is exposed and forms its own facet. (After Smith H I 1974 *Proc. IEEE* **62** 1361.)

such as Ti, Cr or Al (Melliar-Smith and Mogab 1978). However, such an approach is far from desirable in the processing of complex devices.) The alternative, a thicker masking layer, has its own drawbacks. The most serious of these is associated with the phenomenon of *redeposition.* Because sputtering is strictly a physical process, the ejected atoms are readily condensed and, in fact, have a high probability of redepositing on any surface they encounter. The sputtered material is ejected with roughly a cosine distribution and therefore a significant fraction can redeposit on elevated, adjacent surfaces such as mask features. This effect is illustrated schematically in figure 9.

Redeposition not only affects etch bias, but frequently leaves 'ears' (figure 9*d*) after mask removal which can cause considerable difficulty in any subsequent processing. Figure 10 is an SEM illustrating this effect. Redeposition can be minimised in ion milling

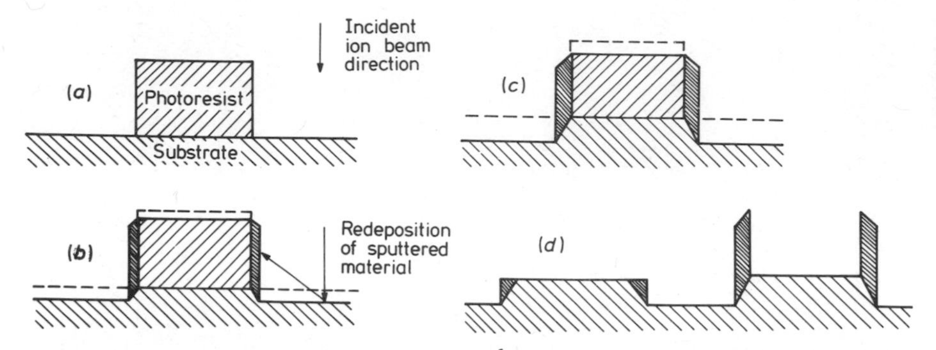

Figure 9. Illustration of the formation of redeposition 'ears'. (*a*) Prior to etching; (*b*) and (*c*) during etching, facets form and sputtered material redeposits on sidewalls; (*d*) after resist stripping the redeposited material may break-off or may remain depending on its thickness and fragility. (After Glöersen P G 1975 *J. Vac. Sci. Technol.* **12** 28.)

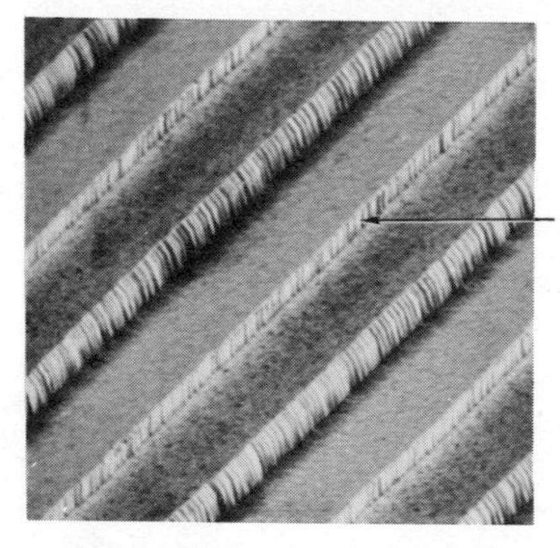

Figure 10. SEM micrograph showing redeposition ears formed during ion beam etching of a 4500 Å thick permalloy film. The photoresist mask has been stripped.

by an appropriate choice of incident angle of the ion beam to provide a higher ion flux to the sidewalls. This approach is not possible with sputter etching where ions arrive at normal incidence to the cathode. Alternatively, the mask can be tapered, but this inevitably impacts on resolution because of sputter erosion of the mask. Other effects such as trenching and a variation of etch rate with surface topography can also influence the resolution of ion etching. For example, the etch rate may be enhanced or suppressed in the vicinity of a step depending on the variation of etch rate with angle of incidence. These effects are covered in Melliar-Smith and Mogab (1978) and in a recent paper by Lee (1979).

In summary, with ion etching, anisotropy is inherent in the directed incidence of energetic, noble ions which are the etchant species. However, the resolution is limited by faceting and redeposition. To achieve optimum resolution a thin, etch resistant mask must be used to minimise redeposition and facet propagation. Generally, this requires processing steps beyond the usual resist delineation. Finally, the selectivity is rather poor. In consequence, ion etching has been employed primarily where patterns having a small ratio of lateral to vertical dimension are to be defined and/or where volatile compounds are unknown. An important example is the fabrication of field access, magnetic bubble memory devices where a high density of micron-size features must be transferred into thin (~ 4500 Å) permalloy (NiFe) films. (There are no known gases which can be used in a reactive plasma mode to create volatile NiFe products.) In contrast, ion etching has found relatively little application to Si integrated circuit fabrication. An exception has been the uniform gold process developed by Labuda *et al* (1974) and Ryden *et al* (1976).

4.2. Plasma and reactive ion etching

Although our understanding of reactive gas plasmas and their interaction with solid surfaces has advanced considerably in the past several years, it is still far from complete. Coburn and Winters (1979a) have recently reviewed the basic phenomena and mechanisms of importance to etching in reactive gas plasmas. Consequently, the present discussion will be limited to a few general comments and observations.

Plasma etching and RIE processes have developed mainly around Si integrated circuit fabrication and therefore the etching of Si, Si compounds and Al metallisation has been at the focal point. Plasmas generated from molecular gases usually containing one or more halogens and often in the form of halocarbons have been used extensively. Table 1 lists some of the gases investigated and materials reportedly etched in plasmas formed from these gases. This list is representative rather than exhaustive.

In general, none of the gases used reacts spontaneously (at practical rates) with any of the materials of interest at the relatively low temperatures required for pattern transfer. Indeed, many of these gases are substantially inert. Once admitted to the discharge, however, they are converted through inelastic collisions with energetic electrons to various fragments of which some, at least, can be highly reactive (Melliar-Smith and Mogab 1978). CF_4, for example, is a very stable and unreactive gas. When introduced into a low pressure discharge, particularly with oxygen present, it dissociates to form F atoms (among other things) which react spontaneously with Si; etch rates of several kÅ min^{-1} can easily be achieved.

CF_x ($x = 1, 2, 3$) radicals and ions, F ions, higher order unsaturated fluorocarbons and free electrons are also present, to some extent, in a CF_4 discharge. Relative concentrations are determined by system parameters such as power, pressure and feed gas flow rate as well as by the materials contained within the reactor. The complexity of the situation is easily appreciated and it is not surprising that a range of phenomena has been observed.

Table 1. Some gases used in plasma etching and RIE.

Etching gas	Etchable materials
CF_4, $CF_4 + O_2$	Si, SiO_2, Si_3N_4, Ti, Mo, Ta, W
C_2F_6, C_3F_8, CHF_3	SiO_2, Si_3N_4
$CClF_3$	Si, Au, Ti
$CBrF_3$	Si, Ti, Pt
CCl_2F_2	GaAs
CCl_4, $CCl_4 + O_2$, $CCl_4 + Cl_2$	Al, Si, Cr
BCl_3, $BCl_3 + Cl_2$	Al

The addition of oxygen results in copious generation of F atoms (Melliar-Smith and Mogab 1978) and allows for highly selective etching of Si relative to SiO_2. However, inasmuch as F atoms react spontaneously with Si and are unaffected by directional electric fields, the etching under these conditions is generally isotropic. This is true whether the plasma etching or RIE mode is used at high frequency (typically 13·56 MHz) (Komiya *et al* 1976, Reinberg 1976, Schwartz *et al* 1979). Recent reports have indicated that anisotropic etching of Si in CF_4–O_2 plasmas is possible at microwave frequency and reduced pressure (Suzuki *et al* 1977, 1979) ($\lesssim$1 mTorr) or reduced frequency (Parry and Rodde 1979) (~8 kHz). It is not known whether F atoms are present in substantial concentrations under such conditions, however. Indeed, it seems likely that for pressures less than about 1 mTorr the oxidation reactions which result in F atom production would be greatly diminished. This is consistent with the very slight maximum increase in etch rate observed when oxygen is added up to 20%, the relatively low absolute values of etch rate attained (~200 Å min^{-1}) despite the comparatively high plasma density (~10^{11} cm^{-3}) achieved at microwave frequency and the observation of undercutting at higher pressure ($\gtrsim$10 mTorr) (Suzuki *et al* 1977, 1979). Similarly, low frequency operation may discourage the generation of F atoms and surely leads to much higher ion energies, other things being equal, as shown recently by Taillet (1979). In any case, these effects again demonstrate the complexity of the phenomena occurring over the rather wide parameter space which is accessible.

Introduction of H_2 or H containing gases to CF_4 plasmas suppresses the F atom concentration by formation of HF and promotes selective etching of SiO_2 relative to Si as first observed by Heinecke (1976) and subsequently confirmed by others (Ephrath 1977, Matsuo and Takehara 1977, Mauer and Carruthers 1979). Partially substituted hydrocarbons such as CHF_3 exhibit similar effects (Heinecke 1976, Lehmann and Widmer 1978, Itoga *et al* 1977, Komiya *et al* 1979). Fluorine 'deficient' ($F/C < 4$) discharges produced from gases such as C_2F_6 or C_3F_8 or by introduction of a large area of F consuming material such as Si to a CF_4 plasma also etch SiO_2 faster than Si (Schwartz *et al* 1979, Heinecke 1976, Matsuo 1978, Mayer 1979). Such conditions can also result in the deposition of fluorocarbon polymers owing to the presence of relatively high concentrations of polymer precursors such as CF_2 radicals (Coburn and Winters 1979a, Coburn and Kay 1979). Thus, rather subtle changes in operating parameters or surface conditions can transform etching into deposition. Neutral CF_x radicals are presumed to be the etchant species for SiO_2, at least for the high pressure conditions used in plasma etching, and generally it is observed that the etching is anisotropic. An explanation of the anisotropy, given that neutrals are the etchant species, has been sought in the influence of charged particle bombardment (Coburn and Winters 1979a). It has been observed that certain combinations of radicals (or ions) and substrates are unreactive or react only slowly in the absence of a plasma. Reaction rates for these species can be greatly enhanced when the substrate is subjected to bombardment by energetic ions or electrons (Coburn and Winters 1979a). Since surfaces immersed in the plasma are bombarded by energetic ions and/or electrons and, moreover, as these particles arrive in a highly directional fashion due to imposed fields, it is clear that when the reaction is ion or electron assisted, anisotropy will result. Possible mechanisms to account for the enhanced reaction rates have been listed by Coburn and Winters (1979a, b); generally they involve utilisation of the energy of the incident charged particles to overcome an activation barrier either for chemisorption, product formation or product desorption.

An alternative explanation would attribute the etching entirely to ionic species which necessarily arrive at the surface with directionality. In the author's opinion the low ion densities ($\sim 10^{10}\,cm^{-3}$) in the plasmas used for etching are difficult to reconcile with the often high etch rates observed (up to several thousand Å min^{-1}), suggesting that this picture is too simplistic. On the other hand, it may be that as pressure is reduced and electron energies increased correspondingly, the concentration of reactive fragments shifts from mostly neutrals to mostly ions. This would be consistent with the fact that anisotropy is often more easily achieved with RIE.

By way of example, it has been reported that single or polycrystalline Si can be reactively ion etched, anisotropically, in Cl_2 plasmas (Schwartz and Schaible 1979). In contrast, Mogab and Levinstein (1979) have observed that Cl_2 plasmas always exhibit undercutting for the etching of polysilicon in a higher pressure, parallel plate plasma etching reactor. The latter investigators have also found that the addition of C_2F_6 to the Cl_2, in sufficient quantity, results in fully anisotropic etching under otherwise identical conditions. They account for these observations on the basis of three effects: (1) Cl atoms, the likely etchant species, react slowly with Si in the absence of energetic particle (ion or electron) bombardment; (2) ion or electron bombardment greatly enhances the reaction rate; and (3) CF_3 radicals can combine with Cl on surfaces to form volatile CF_3Cl; this 'recombination' reaction competes with the etching reaction and is probably suppressed by ion (electron) bombardment. The anisotropy is thought to result because on bombarded surfaces the rate of etching exceeds the rate of recombination while on nonbombarded surfaces the reverse is true. If the gas mixture is adjusted such that there is an abundance of etchant species precursor (Cl_2) relative to recombinant species precursor (C_2F_6), lateral etching is observed.

It is tempting to conclude from the foregoing that in Cl_2 plasmas Si is etched by neutral species (Cl) under plasma etching conditions and by ions (Cl_2^+, Cl^+) under RIE conditions. However, additional evidence again indicates that the situation is more complicated. Fully anisotropic etching of undopred p-type or lightly doped n-type Si is possible using Cl_2 and RIE, but when the Si is heavily doped n-type (N^+) material, significant lateral etching can occur under otherwise identical conditions (Schwartz and Schaible 1979). Moreover, the vertical etch rate for N^+ Si exceeds that of undoped and p-type material under both RIE (Schwartz and Schaible 1979) and plasma etching (Mogab and Levinstein 1979) conditions. In the latter case the rate for N^+ material was found to be about 15 times larger than for undoped Si (Mogab and Levinstein 1979). It has been suggested that this effect is related to the need for charge transfer between semiconductor and etchant species (presumed to be the highly electronegative Cl atom) to effect chemisorption (Mogab and Levinstein 1979). This process can occur more rapidly when a copious supply of free electrons exists in the Si and this, in turn, is reflected in the enhanced etch rate.

The lateral etching of N^+ Si in RIE with Cl_2 implies the existence of some neutral etchant species and leaves open the question of whether ions or neutrals assisted by ions are primarily responsible for etching. It is interesting to note that the addition of sufficient CCl_4 to the Cl_2 prevents lateral attack of N^+ material with RIE (Schwartz and Schaible 1979). Possibly the CCl_4 functions as a recombinant precursor†. CCl_4–Cl_2

† A single component may be *both* etchant and recombinant precursor. For example, electron impact dissociation of CCl_4 yields CCl_3 and Cl. If Cl is the etchant, recombination (that is, $Cl + CCl_3 \rightarrow CCl_4$) will suppress etching.

mixtures also produce anisotropic profiles in Si when using plasma etching (Reinberg 1978).

The foregoing discussion serves to illustrate the complexities inherent in the use of reactive gas plasmas for etching. The physics and chemistry of both plasma and surface phenomena must be understood before a reasonably complete picture can evolve. As yet, no persuasive model has been offered which can account for all the observed phenomena, however, it seems certain that *anisotropic etching* is the result of the interaction of energetic charged particles with the surface undergoing etching, irrespective of whether plasma etching or reactive ion etching conditions are used. The question to be answered is whether, or to what extent, the ions themselves are the etchant species. The answer may differ for RIE as opposed to plasma etching. Substantially more data from well-designed experiments will be needed to answer this question.

The energetic ions which impact surfaces in RIE and plasma etching systems also produce sputtering, of course. Typically, conditions are controlled so that this contributes insignificantly to the etch rate. However, sputtering has another important consequence in reactive plasmas. Any material present which cannot be converted to a volatile product will still be sputtered to some extent. Redeposition of this material will impede etching. Such an effect has been observed in both RIE and plasma etching systems (Vossen 1979, Schwartz *et al* 1979, Ephrath 1978, Bondur 1979, Mogab, unpublished results) and is manifested as surface roughening and residues. The effect is more pronounced in RIE because the lower pressure conditions favour more energetic ions and longer mean free paths for sputtered atoms. The materials internal to the reactor as well as masking materials must be chosen carefully to avoid this problem. Furthermore, even materials of the devices must be considered. For example, it was found that a plasma reactor became contaminated with Au when exposed to Au-metallised devices during removal of a dielectric passivation layer. Subsequent etching of n-type Si wafers in this reactor resulted in Au doping and the formation of a p-skin (Murarka and Mogab 1979).

5. Applications, advantages and limitations

All of the materials commonly found in present day Si integrated circuits can be etched anisotropically with reasonable selectivity in reactive plasmas (Mogab and Harshbarger 1978). Examples of etched features in SiO_2, polycrystalline Si, single crystal Si and Al are shown in figure 11. The methods and processes developed to do this are largely empirical and are not yet entirely trouble-free. Etching of Al metallisation (Poulsen 1977, Reichelderfer *et al* 1977, Schaible *et al* 1978, Heinecke 1978, Schaible and Schwartz 1979) has presented the most difficult problems. Gases like CCl_4 and BCl_3 found to be effective in removing the ever present native oxide from Al are liquids at or near room temperature making them more difficult to handle. Alloy additions such as Cu which do not form volatile chlorides leave residues which often must be 'cleaned-up' by a selective wet chemical process. Chlorine is readily adsorbed on and retained by photoresist. Reaction of this adsorbate with atmospheric moisture leads to formation of HCl and subsequent attack of Al unless the resist is removed *in situ* or other steps are taken to passivate the surfaces prior to atmospheric exposure.

In contrast, high resolution processes for etching SiO_2, phosphosilicate glasses, Si_3N_4 and polycrystalline Si are better developed. Recently, selectivities of 30 : 1 or more,

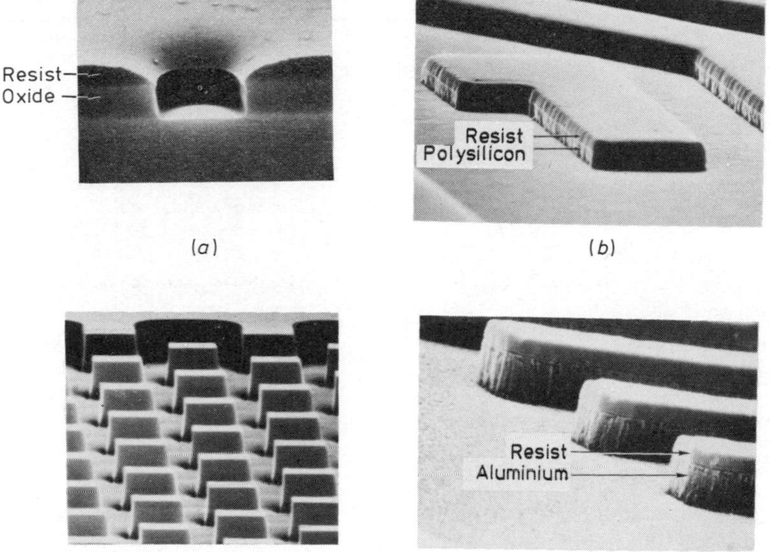

Figure 11. SEM micrographs illustrating anisotropic etching by reactive plasma techniques. (*a*) A 4 μm contact window plasma etched in a 2 μm thick phosphorus doped SiO_2 layer. The substrate is Si. (*b*) Plasma etched pattern in 1 μm thick phosphorus doped polycrystalline Si film. The substrate is SiO_2. (*c*) 1 μm wide features created by RIE of single crystal Si. (Courtesy of D N K Wang and D Maydan.) (*d*) Plasma etched pattern in a 1·5 μm thick Al-0·7% Cu film. The substrate is a plasma deposited SiO_2 film. (Courtesty of A D Butherus.)

with respect to underlying materials, have been achieved in typical applications such as etching of contact windows in an intermediate insulator or polysilicon gate and interconnect definition. These values are ample for present SIC designs particularly in conjunction with processing schemes which reduce vertical topography.

When properly designed, the reactive plasma techniques provide photoresist compatible, high resolution etching, free from redeposition and with good selectivity. In consequence, these will continue to be the techniques of choice for SIC fabrication by subtractive patterning and it is unlikely that inert ion etching will ever be a serious contender despite any future refinements. On the other hand, for materials for which no suitable reactive gas can be found, ion etching methods will be necessary.

An often asked question bears on the efficacy of anisotropic plasma etching relative to reactive ion etching: 'What is the best process?' No clear trend is evident as yet and none may develop since both approaches have been reasonably successful. Nevertheless, it is useful to compare these techniques with respect to practical considerations. An attempt at this is made in table 2 based on information available in the literature and the author's own experience. A danger in devising such a comparison is that generalisations to which specific exceptions can always be found must be made. Further, the rapid evolution of the art may produce significant changes in the near future.

A few comments on table 2 are in order. The higher pressure operation characterising plasma etching implies reduced particle energy and better heat transfer. This makes for

Table 2. Comparison of anisotropic plasma etching and RIE.†

Parameter	Plasma etch	RIE	Comments
Selectivity with respect to:			
(*a*) resist	+	−	
(*b*) substrate	0	0	Depends on film/substrate combination
Anisotropy	−, 0	+, 0	Depends on material
Etch rate	+	−	Not remarkably different
Uniformity	0	0	
Radiation damage	+	−	Depends on device level, generally anneable
Redeposition of involatile materials	+	−	Can usually be avoided by proper choice of construction and masking materials
Equipment cost/complexity	+	−	

† + = better than; − = worse than; 0 = equivalent.

a less harsh environment for organic resists. The lower energy also means less tendency to radiation damage and redeposition of involatile material due to sputtering. On the other hand, anisotropy is less easily achieved at comparable selectivity for certain film–substrate combinations. Finally, higher pressure operation usually means a higher concentration of active species resulting in somewhat higher etch rates and less sensitivity to residual vacuum levels. The latter factor accounts for the reduced complexity and cost of equipment for plasma etching as mechanical pumps are usually sufficient. RIE requires high vacuum pumping capability.

Radiation damage due to plasma exposure is dependent on both the device structure and the plasma environment. MOS devices are known to be quite sensitive to this type of damage when the gate oxide is exposed directly (Melliar-Smith and Mogab 1978). Studies of oxygen plasma stripping of photoresists from gate oxides (Van DeVen and Kalter 1976, Maddox and Parker 1978) and of RIE of oxide films in CF_4 (DiMaria *et al* 1978) have revealed deleterious effects such as mobile ion contamination and charge trapping. However, various techniques were found effective in gettering or immobilising contaminant ions (Maddox and Parker 1978, Kalter and Van DeVen 1976) and traps were annealed out at 600°C in 30 min (DiMaria *et al* 1978). Further, for pattern transfer the gate oxide is seldom exposed directly. Usually it is covered by, at least, the gate electrode material which affords protection against ion, electron and near UV photon damage. The primary concern here is soft x-rays which have much greater penetration power. They originate from bremsstrahlung processes involving high energy secondary electrons. The minimum x-ray wavelength is determined by the peak-to-peak voltage applied to the driven electrode. In general, this voltage can be kept sufficiently low with plasma etching and RIE that damage either does not occur or it can be annealed out during subsequent processing. Paradoxically, a recent report indicates that fixed charge and surface states can be annealed out of MOS structures by exposing them to low pressure RF plasmas run under certain, rather exacting, conditions (Ma and Ma 1978).

Sputter etching can be more of a problem in this regard since relatively high voltages are required for practical etch rates. A method of suppressing x-ray impingement on the substrates has been devised to circumvent this problem (Ryden *et al* 1976).

Plasma-assisted etching methods, in general, are not very compatible with present day electron beam and x-ray resists (Bowden and Thompson 1979). This has spawned multi-level approaches which allow full advantage to be taken of the high resolution lithographies which require such resists. A particularly promising example of this approach was described recently by Moran and Maydan (1979). A thick plasma-tolerant polymeric material is spun-on the substrate and overcoated with a thin intermediate layer such as SiO_2 and finally by the working resist. A pattern is generated in the resist, lithographically, and then transferred into the SiO_2 by anisotropic plasma etching or RIE. The SiO_2 is made sufficiently thin (~1000 Å) that the intolerant resist mask need be exposed to the plasma for only a brief period. Finally, the SiO_2 is used as a mask for ansitropic reactive ion etching of the thick polymer which subsequently serves as a mask for the final pattern transfer to the substrate. This so-called *tri-level process* is illustrated schematically in figure 12. Note that the thick spun-on bottom layer tends to produce a planar surface for the final resist layer despite the existence of stepped topography on the substrate. This permits the use of a thin resist layer for maximum resolution. Clearly, this process has added complexity, but at present a multi-level approach appears to be the only viable alternative for plasma patterning at the working limits of these advanced lithographies.

Reports of the application of reactive plasma patterning methods in non-Si based technologies have been sparse, to date. Recently, plasma etching of Al–Cu metallisation has been employed in the fabrication of isoplanar field access bubble devices (Bullock *et al* 1979) as well as to a new generation of conductor access bubble memories (Bobeck *et al* 1979). A combination of ion milling and plasma etching has been used to etch fine gratings in GaAs (Somekh *et al* 1976) and there is currently substantial interest in the development of plasma processing for GaAs and other III–V compound devices.

Other advantages and limitations of plasma-assisted etching methods and the inevitable comparisons to liquid etching have been covered previously (Melliar-Smith and Mogab 1978) and will not be repeated. It is appropriate, however, to discuss briefly the use of end-point detection.

Various means of end-point detection have been explored. They include: (1) direct visual observation of the etched layer; (2) monitoring of optical reflections from the etched layer (Busta *et al* 1978); (3) detection of changes in the concentration of etch

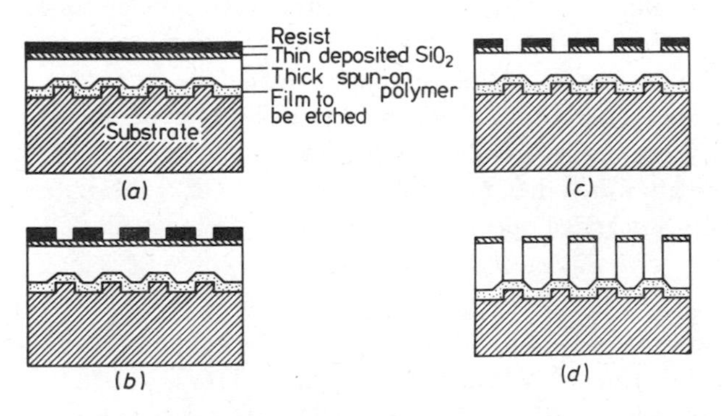

Figure 12. Schematic illustration of the tri-level mask process of Moran and Maydan (1979). (*a*) Ready for patterning; (*b*) resist patterned by lithography; (*c*) oxide patterned by anisotropic etching; (*d*) polymer patterned by anisotropic etching, ready for final pattern transfer. Note that the stepped topography is planarised by the spun-on polymer layer.

active species in the plasma by emission spectroscopy (Harshbarger *et al* 1977, Mogab 1977, Harshbarger and Porter 1978, Greene 1978); (4) detection of etch products by emission spectroscopy (Bernstein and Labuda 1973, Degenkolb *et al* 1976, Griffiths and Degenkolb 1977, Curtis and Brunner 1978) or mass spectrometry (Raby 1978, Oshima 1978, Hosaka *et al* 1979); and (5) detection of changes in plasma impedance (Ukai and Hanazawa 1979). Methods (1) and (2) are independent of the area of material being etched, but are not well suited to dealing with nonuniform batch ending. Methods (3)–(5) require a minimal area of material determined by etch rate and detector sensitivity and tend to average over nonuniformities.

Because film thickness and etch rate uniformity is never perfect, end-point detection should be viewed as a useful but *approximate* diagnostic. With this in mind, plasma-assisted pattern transfer processes should be designed with sufficient anisotropy and selectivity to permit overetching rather than being critically dependent upon end-point detection *for linewidth control.* (Recall that overetching is required in any case when anisotropic etching is employed to remove material from stepped surfaces.) End-point detection when coupled with well-designed processes provides a very convenient means for compensating for variations in etch rate resulting from fluctuations in material composition or thickness or from changes in operating parameters.

6. Conclusion

Ion, plasma and reactive ion etching techniques are versatile means for implementing high resolution pattern transfer as required in the fabrication of present day electron devices. Inert ion etching methods are resolution limited by faceting and redeposition effects and are not very selective, but they have the virtues of being applicable to any material and of being relatively well understood. Plasma etching and RIE methods rely on chemical reaction to produce volatile compounds and so can be made relatively selective, but are limited to materials which can be readily converted to volatile compounds at or near room temperature. Under suitable conditions the chemical reactions can be driven by charged particle bombardment leading to highly anisotropic etching with minimal physical sputtering. The resolution of pattern transfer is then virtually equivalent to that of the lithographic process used to generate the etch mask.

The reactive plasma methods will likely be the techniques of choice for subtractive patterning of fine features so long as suitably volatile compounds can be found. Inert ion etching will continue to be useful for patterning other materials.

It seems safe to predict that the future will bring much improved plasma processes with respect to selectivity, uniformity, reproducibility and throughput for SIC fabrication as well as increased utilisation in other technologies such as GaAs integrated circuits, magnetic bubble memories, integrated optics and advanced display systems.

Acknowledgments

The author is grateful to W A Johnson for supplying the data of figure 7, to D N K Wang and D Maydan for figure 11(*c*) and to A D Butherus for figure 11(*d*).

References

Bernstein T and Labuda E F 1973 *J. Vac. Sci. Technol.* **10** 108
Bobeck A H, Blank S L, Butherus A D, Ciak F J and Strauss W 1979 *Bell Syst. Tech. J.* **58** 1453

Bondur J A 1979 *J. Electrochem. Soc.* **126** 226
Bowden M J and Thompson L F 1979 *Solid-St. Technol.* **22** 72
Bullock D C, Fontana R E Jr, Singh S K, Bush M and Stein R 1979 *IEEE Trans. Magnetics.* **MAG-15** 1697
Busta H H, Lajos R E and Kiewit D A 1978 *Industrial Res. Dev.* **20 (6)** 133
Coburn J W and Kay E 1972 *J. Appl. Phys.* **43** 4965
—— 1979 *IBM J. Res. Dev.* **23** 33
Coburn J W and Winters H F 1979a *J. Vac. Sci. Technol.* **16** 391
—— 1979b *J. Appl. Phys.* **50** 3189
Curtis B J and Brunner H J 1978 *J. Electrochem. Soc.* **125** 829
Degenkolb E O, Mogab C J, Goldrick M R and Griffiths J E 1976 *Appl. Spectrosc.* **30** 520
DiMaria D J, Ephrath L M and Young D R 1978 *paper presented at 20th Conf. on Electronic Materials, Santa Barbara, CA, June 1978*
Ephrath L M 1977 *Electrochem. Soc. Extended Abstract* **77–2** 376
—— 1978 *J. Electron. Mater.* **7** 415
Ephrath L M and Mogab C J 1979 *Report on Microstructure Science, Engineering and Technology* (Washington DC: National Academy of Sciences) pp 6.2–6.14
Greene J E 1978 *J. Vac. Sci. Technol.* **15** 1718
Griffiths J E and Degenkolb E O 1977 *Appl. Spectrosc.* **31** 134
Harshbarger W R and Porter R A 1978 *Solid-St. Technol.* **21** 99
Harshbarger W R, Miller T A, Norton P and Porter R A 1977 *Appl. Spectrosc.* **31** 201
Heinecke R A H 1976 *Solid-St. Electron.* **19** 1039
—— 1978 *Solid-St. Technol.* **21** 104
Hosaka S, Sakudo N and Hashimoto S 1979 *J. Vac. Sci. Technol.* **16** 913
Itoga M, Inoue M, Kitahara Y and Ban Y 1977 *Electrochem. Soc. Extended Abstract* **77–2** 378
Kalter H and Van DeVen E P G T 1976 *Electrochem. Soc. Extended Abstract* **76–1** 335
Koenig H R and Maissel L I 1970 *IBM J. Res. Dev.* **14** 168
Komiya H, Toyoda H, Kato T and Inaba K 1976 *Suppl. Jap. J. Appl. Phys.* **15** 19
Komiya H, Toyoda H and Itakura H 1979 *paper presented at 21st Conf. on Electronic Materials, Boulder, Colorado, June 1979*
Labuda E F, Herb G K, Ryden W D, Fritzinger L B and Szabo J M Jr 1974 *Electrochem. Soc. Extended Abstract* **74–1** 195
Lee R E 1979 *J. Vac. Sci. Technol.* **16** 164
Lehmann H W and Widmer R 1978 *J. Vac. Sci. Technol.* **15** 319
Ma T-P and Ma W H-L 1978 *Appl. Phys. Lett.* **32** 441
Maddox R L and Parker H L 1978 *Solid-St. Technol.* **21** 107
Matsuo S 1978 *Jap. J. Appl. Phys.* **17** 235
Matsuo S and Takehara Y 1977 *Jap. J. Appl. Phys.* **16** 175
Mauer J L and Carruthers R A 1979 *paper presented at 21st Conf. on Electronic Materials, Boulder, Colorado, June 1979*
Mayer T M 1979 *paper presented at 21st Conf. on Electronic Materials, Boulder, Colorado, June 1979*
Melliar-Smith C M 1976 *J. Vac. Sci. Technol.* **13** 1008
Melliar-Smith C M and Mogab C J 1978 *Thin Film Processes* ed J L Vossen and W Kern (New York: Academic Press) Ch. V-2
Mogab C J 1977 *J. Electrochem. Soc.* **124** 1262
Mogab C J and Harshbarger W R 1978 *Electronics* **115** 117
Mogab C J and Levinstein H J 1979 *J. Vac. Sci. Technol.* **16**
Moran J M and Maydan D 1979 *Bell Syst. Tech. J.* **58** 1027
Murarka S P and Mogab C J 1979 *J. Electron. Mater.* **8** 763
Oshima M 1978 *Jap. J. Appl. Phys.* **17** 579
Parry P D and Rodde A F 1979 *Solid-St. Technol.* **22** 125
Poulsen R G 1977 *J. Vac. Sci. Technol.* **14** 266
Raby B A 1978 *J. Vac. Sci. Technol.* **15** 205
Reichelderfer R, Vogel D and Bersin R L 1977 *Electrochem. Soc. Extended Abstract* **77–2** 414
Reinberg A R 1976 *Etching for Pattern Definition* ed H G Hughes and M J Rand (Princeton, New Jersey: Electrochemical Society) p 91

—— 1978 *US Patent No.* 4 069 096
Ryden W D, Labuda E F and Clemens J T 1976 *Etching for Pattern Definition* ed H G Hughes and M J Rand (Princeton, New Jersey: Electrochemical Society) p 144
Schaible P M, Metzger W C and Anderson J P 1978 *J. Vac. Sci. Technol.* **15** 334
Schaible P M and Schwartz G C 1979 *J. Vac. Sci. Technol.* **16** 377
Schwartz G C, Rothman L B and Schopen T J 1979 *J. Electrochem. Soc.* **126** 464
Schwartz G C and Schaible P M 1979 *J. Vac. Sci. Technol.* **16** 410
Somekh S, Casey H C and Ilegems M 1976 *Appl. Opt.* **15** 1905
Suzuki K, Okudaira S and Kanomata I 1979 *J. Electrochem. Soc.* **126** 1024
Suzuki K, Okudaira S, Sakudo N and Kanomata I 1977 *Jap. J. Appl. Phys.* **16** 1979
Taillet J 1979 *J. Physique* **11** L-223
Ukai K and Hanazawa K 1979 *J. Vac. Sci. Technol.* **16** 385
Van DeVen E P G T and Kalter H 1976 *Electrochem. Soc. Extended Abstract* **76-1** 332
Vossen J L 1979 *J. Electrochem. Soc.* **126** 319
Vossen J L and Cuomo J J 1978 *Thin Film Processes* ed J L Vossen and W Kern (New York: Academic Press) Ch. II–I

Solar cells

H Fischer† and K Roy
AEG-Telefunken, Semiconductor Division, Heilbronn, West Germany

Abstract. Direct conversion of sunlight into electrical energy is realised with solar cells, a semiconductor device and a converter component which presently exhibits reasonable efficiency. For power generation in space applications, single-crystalline Si solar cells have dominated the field for more than a decade. Large-area flexible solar generators in the 10 kW power range require advanced cell designs which approach the limitations of Si technology. Current development trends can be outlined as: increase in conversion efficiency and its limitation by component and materials properties; reduction in operation temperature in thermal equilibrium; and increase in power–mass ratio by using thin cells with multiple light passes.

To make solar cells viable for terrestrial power generation, their fabrication costs must be reduced from some 10 DM/W to 1 DM/W. Therefore, instead of semiconductor-grade single-crystalline Si, low-grade Si has to be used and solar cell structures have to be applied which tolerate a high degree of crystalline imperfections up to polycrystalline material. Various approaches currently under investigation are reported, especially the state of the art in non-single-crystalline cells. High-output production methods will be described, as well as experience in pilot production lines. Based on the state of technology reached and present production experience, it can be concluded that solar cells will be important components in terrestrial energy conversion in the next decade.

1. Introduction

Direct conversion of sunlight into electrical energy is realised with the solar cell, a semiconductor device, the only converter component which presently exhibits a reasonable conversion efficiency. For space applications, single-crystalline Si solar cells have dominated the field of power generators for more than a decade. Si solar cells are also extremely attractive candidates for large-scale terrestrial power generation, because they have desirable conversion efficiencies, the starting materials for Si production are abundantly available, and Si manufacturing technology and the device industry are highly developed.

Since 1973 the impact of the oil crisis has expanded solar cell research into a worldwide effort considering numerous materials and structures. Only Si cells have yet reached the state of industrial realisation. This paper will therefore concentrate on Si solar cells, their fundamentals, material and device characteristics, and the state of the art in development trends toward increased performance and lower fabrication costs.

† Present address: Robert Bosch GmbH, Semiconductor Division, Tübinger Strasse 123, 7410 Reutlingen 1, West Germany.

0305-2346/80/0053-0055 $03.00

2. Fundamentals and basic characteristics

The physics and technology of solar cells have been treated in various books (Hovel 1975) and review articles (Fischer 1974, Bucher 1978, IEEE Photovoltaic Specialist Conferences), and is only briefly discussed here. The conversion of light into electrical energy in a solar cell is caused by the photovoltaic effect which appears at boundary layers of a semiconductor as a result of the absorption of light. Electrons and holes which are generated by absorption processes in excess of thermal equilibrium values can be separated at the internal electric potential which exists across the space charge region of the boundary layer. This causes a reduction in the internal potential, and the difference from the thermal equilibrium value appears as a photovoltage, which can drive a photocurrent through an external circuit.

The basic structure of a typical cell is shown in figure 1. The solar cell consists of a p-type Si wafer. Under the front side a shallow p–n junction is formed by indiffusion of P. Electrical contacts are achieved by evaporated metal layers, which completely cover the rear side. The grid structure on the front side minimises the sheet resistance of the diffused layer. An anti-reflex cover gives the cell surface a blue appearance.

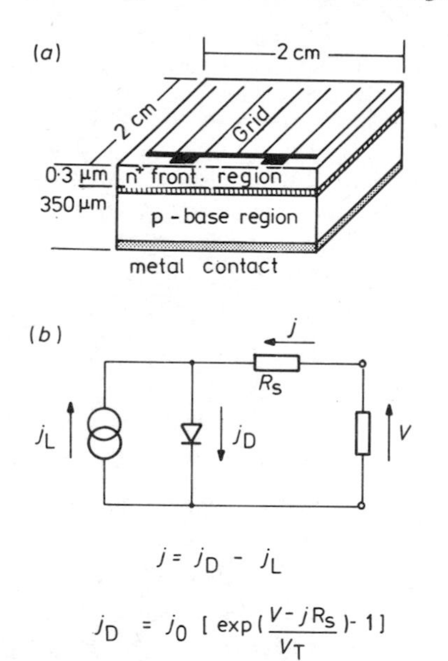

Figure 1. (*a*) Schematic cross section and (*b*) equivalent circuit of a solar cell.

2.1. I–V characteristics

The simplest equivalent circuit is given in figure 1(*b*), and consists of a constant current source, which produces the light-generated current j_L. This current is then separated in an external load current and an internal diode current, which exhibits a rectifier characteristic. Figure 2 displays the I–V characteristic of an illuminated solar cell, where I_{sc} = short-circuit current which is equivalent to the light-generated current; V_{oc} = open-circuit voltage; P_m = maximum extractable electrical power; CF = curve factor, its deviation from the ideal value (approximately 0·82) characterises the quality of the technology. Figure 2(*b*) gives a more general display of a I–V characteristic for a practical Si solar cell.

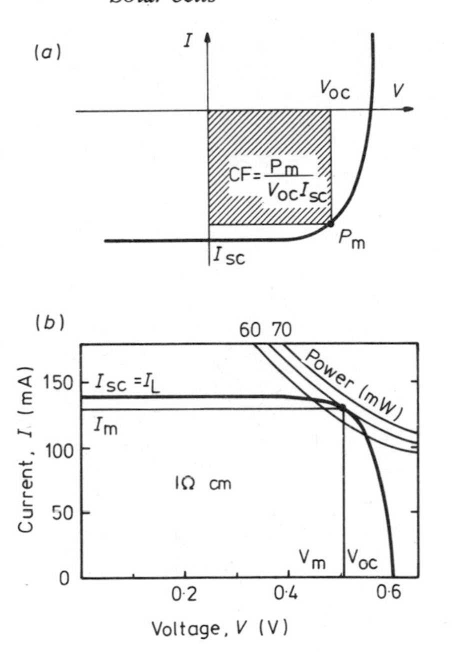

Figure 2. Current–voltage characteristic of a photovoltaic cell and of a standard 1 Ω cm Si solar cell (2 cm × 2 cm).

2.2. *Light-generated current*

The light-generated current is given by

$$j_L = \frac{q}{h \cdot c} \int_0^{\lambda_G} Q(\lambda) H(\lambda) \lambda \, d\lambda \qquad (\text{mA cm}^{-2}), \tag{1}$$

where $H(\lambda)$ is the spectral irradiance of the incident light, λ_G is the absorption edge of the semiconductor ($\lambda_G = hc/E_G$), and $Q(\lambda)$ is the collection efficiency. This term characterises the ratio of photogenerated carriers which are collected at the junction and total carriers produced by absorption. Figure 3 displays a simple model to derive $Q(\lambda)$. The junction depth is x_j, and L_p and L_n are the minority carrier diffusion lengths in the n- and p-regions, respectively. Incident photons were partly absorbed in the n^+ skin under the surface but mainly in the base layer. Only carriers which can diffuse to the junction can contribute to the current, the others are lost due to bulk or surface recombination. Thus $Q(\lambda)$ is determined by configuration, diffusion length of minority carriers L_n and L_p, and surface recombination terms, and can be derived by solving the appropriate transport equations (Hovel 1975). Figure 3 displays the collection efficiency as a function of wavelength. Contributions of the skin and base regions are plotted separately and indicate that the blue part of the spectrum is used in the front layer and the red part in the base of the solar cell.

An ideal cell is characterised by $Q = 1$ for $\lambda \leqslant \lambda_G$, and $Q = 0$ for $\lambda \geqslant \lambda_G$. Figure 4 gives the ideal current as a function of cell thickness for various insolations. The meaning of AM0 to AM2 is given in table 1. Ideal current density ranges from about 30 to 50 mA cm^{-2} under these conditions for a 100 μm thick cell. Practical values for Si solar cells range from about 85 to 90% of these values, including residual reflection and grid pattern losses.

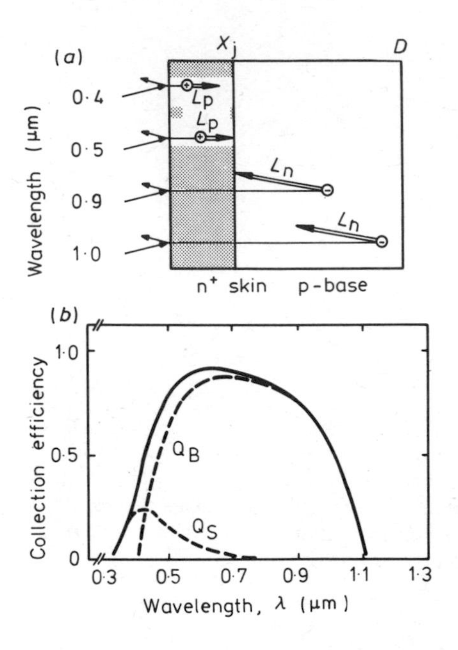

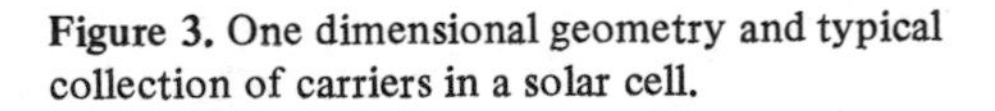
Figure 3. One dimensional geometry and typical collection of carriers in a solar cell.

As can be noted from figure 4, the photocurrent does not depend linearly on incident light power, but is strongly influenced by the spectral distribution of the light. This problem gives rise to numerous misleading results in photovoltaic measurements and calibrations, especially in efficiency data. In general, the efficiency of the same cell increases with increasing AM number. Figure 5 displays as an example the calibration factor AM0/AM1 as a function of cell thickness. While the light power ratio AM0/AM1 is 1·35, the light current ratio is around 1·23, thus giving rise to an increase in conversion efficiency by a factor of about 10%.

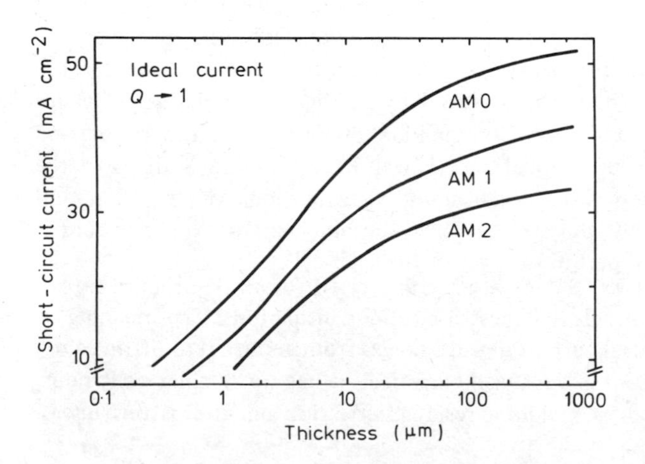

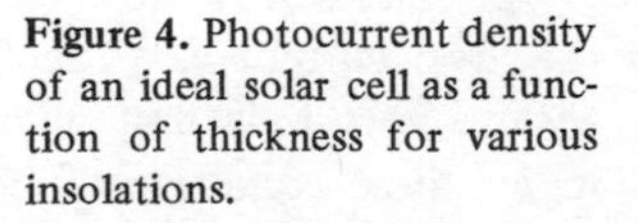
Figure 4. Photocurrent density of an ideal solar cell as a function of thickness for various insolations.

Table 1. The air mass number is equal to the amount of atmosphere the sunlight must penetrate before striking the solar cell.

Air mass	Depth of atmosphere	Solar power ($W\,m^{-2}$)
0	Zero (in space)	1353
1	1 (sea level, sun at zenith)	1000
2	2 (sea level, sun 60° from zenith)	750
3	3 (sea level, sun 72·5° from zenith)	620

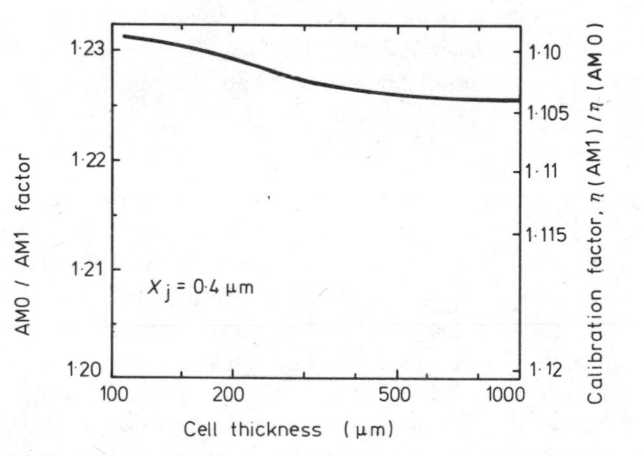

Figure 5. Calibration factor AM0/AM1 as function of cell thickness (AM0/AM1 = 1·35).

2.3. *Open circuit voltage*

The open-circuit voltage is given by

$$V_{oc} = V_T \ln \left[(j_L/j_0) + 1\right], \tag{2}$$

where $V_T = 25{\cdot}8$ mV at 28 °C, and j_0 is the ideal p–n junction saturation current.

2.4. *Maximum power and efficiency*

Maximum power is given by

$$P_m = \text{CF} \times I_{sc} \times V_{oc}, \tag{3}$$

where the ideal curve factor CF (neglecting resistance losses) is given by the following equations

$$V_m = V_{oc} - V_T \ln \left[(V_m/V_T) + 1\right] \tag{4a}$$

$$\text{CF} = \frac{V_m^2}{(V_T + V_m)\, V_{oc}}. \tag{4b}$$

The efficiency of the energy conversion results then in

$$\eta = \frac{\text{maximum electrical power out}}{\text{solar power in}} = \frac{P_m}{P_L}. \tag{5}$$

These phenomenological relations describe completely the operation of a solar cell. They contain basically two parameters j_L and j_0 which depend on the material parameters and incident radiation. Highest maximum power is achieved if the photogenerated current j_L is high, and the diode saturation current j_0 is as low as possible.

3. Technological optimisation of solar cell performance

In contrast to the ideal cell, as described in § 2, the practical solar cell incorporates some power loss factors which, in principle, can be made as small as desired, but not in practice because of limitations of material parameters and technology. Thus improving the performance of solar cells means setting optimum compromises between conflicting parameters.

3.1. Increased blue response

In contrast to the theoretical value presented in figure 3, in a normal cell only about 20% of the available 'blue–violet' photons are used for current generation. The problem is illustrated in figure 6(*a*), which shows a typical diffusion profile of a standard cell (full line), which is the well known anomalous phosphorus profile. The constant impurity zone underneath the surface, given by the solid solubility level, behaves as a degenerated semiconductor with practically zero carrier lifetime. No current contribution will be expected from that 'dead layer' (Lindmayer and Allison 1973). Low-temperature diffusion with controlled surface concentration creates an ideal diffusion profile (broken line).

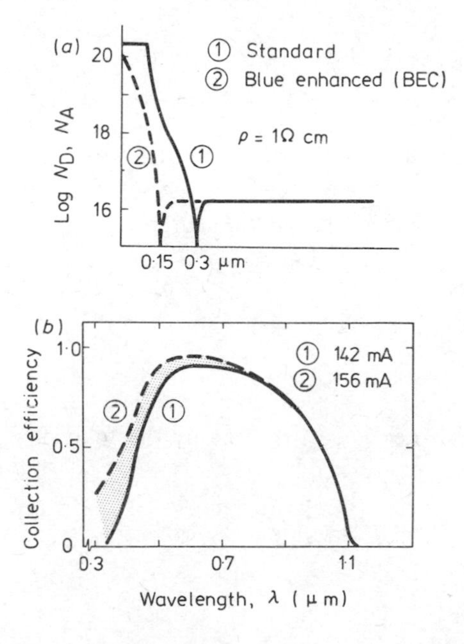

Figure 6. Diffused doping distribution profile and associated collection efficiency of a standard and a blue enhanced cell.

Since a strong electric drift field exists over the diffused region, which forces the minority carrier towards the junction, recombination losses can be neglected. The increase in the blue response is shown in figure 6(*b*). Nearly ideal collection on the blue side of the spectrum is achieved.

3.2. Controlled carrier lifetime and maximum efficiency

As indicated in figure 3, the magnitude of the carrier lifetime is the key parameter for complete collection (Graff and Fischer 1979). Thus an increase in the carrier lifetime will increase the light-generated current. An increase in doping level in the base will reduce the saturation current and so increase the photovoltage (figure 2), while simultaneously the carrier lifetime has to be maintained. Figure 7 shows the AM0 efficiency as a function of diffusion length and carrier lifetime. Optimum conditions are achieved if the diffusion length approaches the cell thickness. A further increase appears if a BSF contact is applied (see §3.3). It should be noted that carrier lifetimes exceeding 10 μs (preferably up to 100 μs) are important for efficient solar cells.

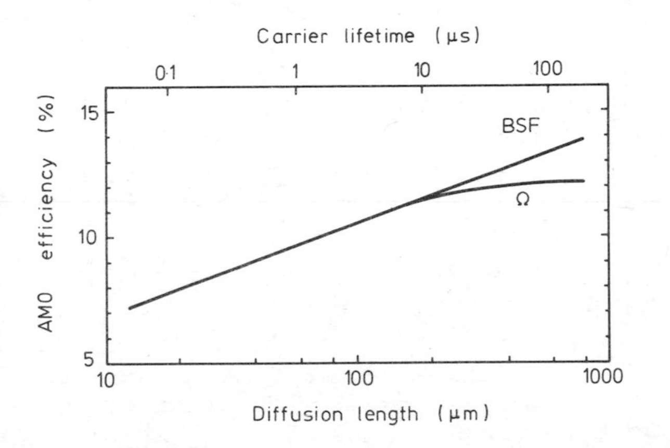

Figure 7. AM0 efficiency as function of diffusion length and carrier lifetime for a cell with an ohmic rear side contact (Ω) and a back surface field contact (BSF).

The carrier lifetime in Si at present is determined by efficient non-radiative recombination, caused by the different impurity contents of the crystals. A dependence of carrier lifetime on carrier concentration in p-type Si is not observed before the specific resistance is below 0·5 Ω cm, as shown in figure 8. In this region, Auger recombination becomes effective (Beck and Conrad 1973). Figure 8 shows measured values of carrier lifetimes for various Si crystals. Measurement points corresponding to different growing techniques are marked. Note that FZ Si exhibits a value approximately 1 magnitude higher than CZ Si, and in the low resistance range only FZ Si can be used for efficient solar cells. In practice, the initial lifetime in the as-grown crystal is changed by process-dependent effects (Graff and Fischer 1979). The crystal undergoes several temperature treatments during the preparation of the device, and the quantity of impurities that diffuse into the crystal will increase. During the operation, some of them can be compensated by getter mechanisms. The influence of the cleanness of the sample surface and of the furnace tube is shown in figure 9. An increase in the carrier lifetime by almost a factor of ten can be achieved by proper control of diffusion and subsequent annealing and gettering conditions.

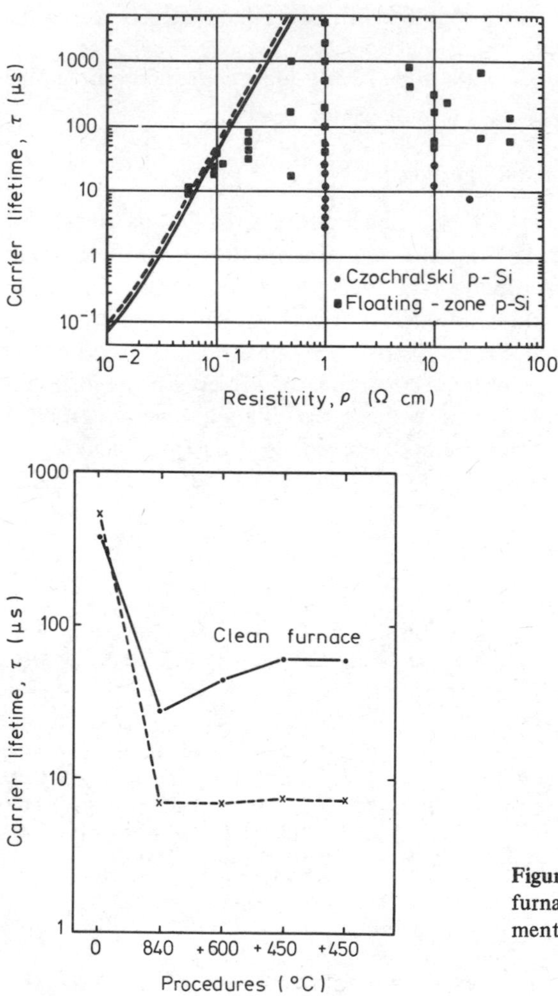

Figure 8. Carrier lifetime as a function of the resistivity in various Si crystals.

Figure 9. Influence of the cleanness of the furnace on the carrier lifetime after heat treatment.

3.3. Thin solar cells with increased performance

With the increasing area of space solar generators, the mass of the individual solar cell is of increasing importance. Figure 10 shows a picture of the space telescope satellite equipped with a flexible roll-out solar generator of 52 m^2 covered with more than 50 000 individual solar cells. To increase the power per mass ratio of the generator, high-efficiency cells with decreasing thickness are required. With the introduction of a back surface field (BSF) at the rear contact area that problem can be solved. Figure 11 shows the mode of operation of that technology. Minority carrier concentration profiles of illuminated cells with ohmic (Ω) and BSF contacts are indicated. Due to the high surface recombination velocity associated with an ohmic contact, the effective carrier lifetime, and also the overall collection efficiency of photogenerated carriers is reduced with D^2. For a 100 μm cell, therefore, the effective carrier lifetime is 1·5 μs, independent of material quality. Simultaneously, the saturation current j_0 increases in proportion to $1/D$,

Figure 10. Space telescope satellite.

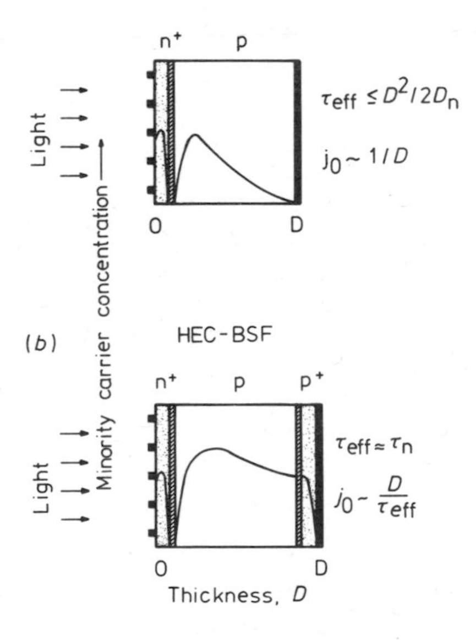

Figure 11. Minority carrier profiles of illuminated n-on-p solar cells under short-circuit conditions.

and so the open-circuit voltage is reduced. Figure 12 shows the strong impact of cell thickness on conversion efficiency for a cell with an ohmic contact. In the case of a BSF contact, which essentially consists of a p^+ region created at the rear side underneath the contact (see figure 11*b*), the minority carriers are reflected at the rear side instead to recombine. As a consequence, the effective carrier lifetime is equal to the specific volume value, which leads to more efficient collection of carriers with reduced thicknesses. If carrier lifetime is kept constant the saturation current remains constant or even decreases with D, and so the open-circuit voltage increases. Figure 12 shows how efficiency is kept constant with decreasing thickness until it drops below 100 μm because of fewer absorbed photons. The impact of that technology is expected to increase the power to mass ratio from the present 0·2 kW kg^{-1} up to values exceeding 1·5 kW kg^{-1} in the future (figure 13).

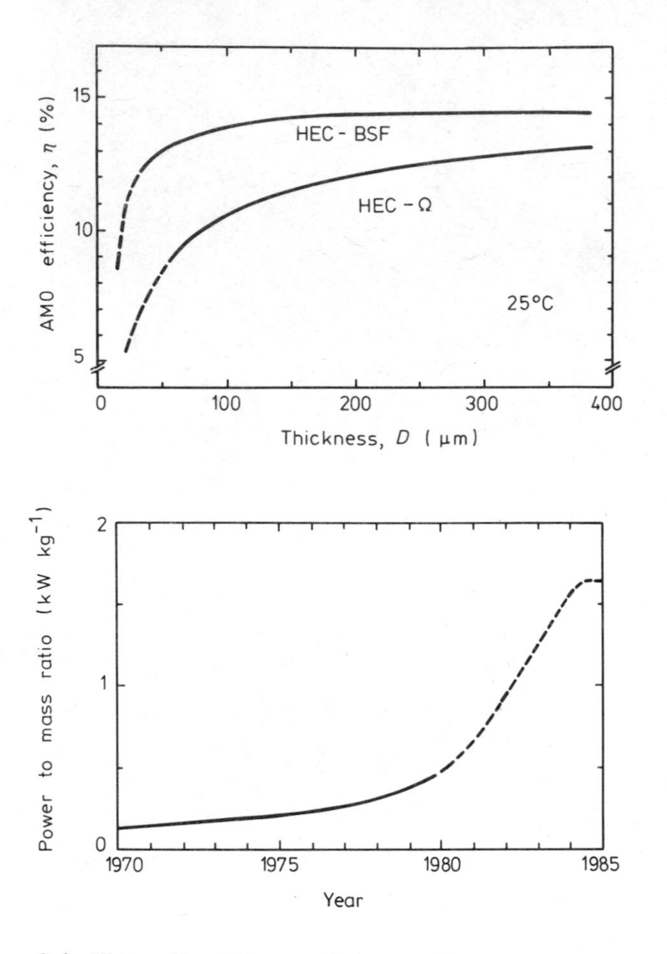

Figure 12. Typical efficiency of a normal high-efficiency cell (HEC) and a high-efficiency cell with back surface field (HEC–BSF) as function of thickness.

Figure 13. Power to mass ratio of Si solar cells from 1970 and expected values up to 1985.

3.4. *Thin cells with controlled operation temperatures*

Solar cells have conversion efficiencies lower than unity, and most of the absorbed energy is converted into heat. As a consequence, the operation temperature increases to about 50 °C in space operation. About 24% of the Sun's energy falls in a spectral region below the band gap of Si. This energy is not absorbed in the Si but at the rear contact of a solar cell, as indicated in spectral reflectance measurements shown in figure 14. The introduction of a reflecting metal layer (back surface reflector) between the rear side and the contact leads to a complete reflection of the IR light through the cell surface, and so to a reduced overall optical absorbance and reduced operation temperature (Rasch *et al* 1979). Figure 15 shows the effect of a BSR on the conversion efficiency in comparison with various other cell types.

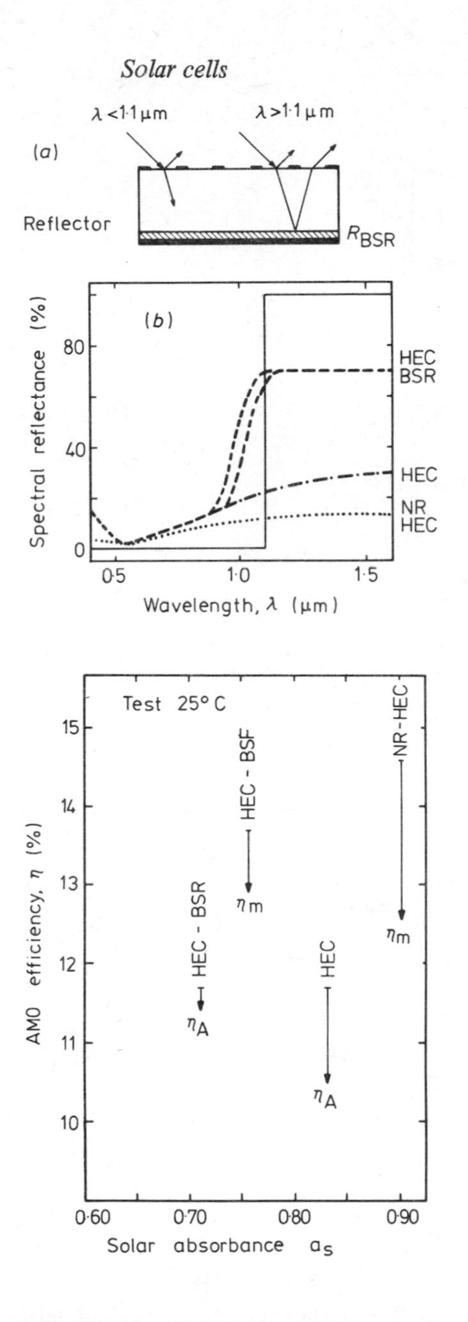

Figure 14. (*a*) Solar cell with back surface reflector (BSR). (*b*) typical spectral reflectance for various Si solar cells.

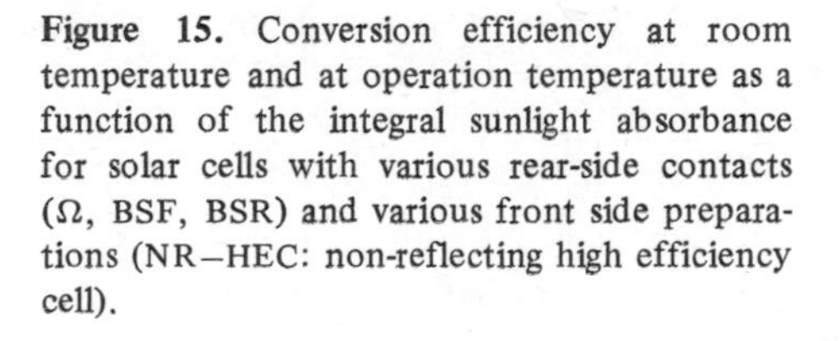

Figure 15. Conversion efficiency at room temperature and at operation temperature as a function of the integral sunlight absorbance for solar cells with various rear-side contacts (Ω, BSF, BSR) and various front side preparations (NR–HEC: non-reflecting high efficiency cell).

3.5. Ultimate conversion efficiency

Increasing knowledge of material parameters, their process dependence and the increased capability of the technology have recently led to solar cells with increased efficiency. Fundamental limitations have been discovered, such as Auger recombination (Fischer and Pschunder 1975). Limitation of the saturation current exists due to the reverse injection of carriers from the base into the diffused zone, and from an increase in recombination which appears as a result of band gap shrinkage at high doping levels (Dunbar and Hauser 1977). Figure 16 displays the maximum achievable AM0 efficiency as a func-

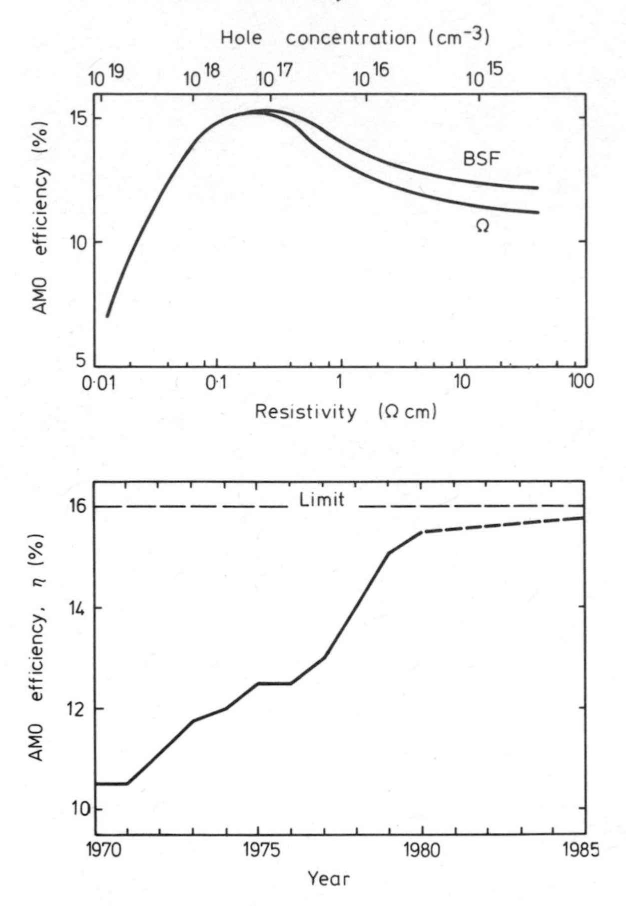

Figure 16. Conversion efficiency as a function of resistivity for Si solar cells. Diffusion length equal thickness, $t = 350\ \mu m$.

Figure 17. Increase in AM0 efficiency since 1970.

tion of resistivity. A maximum of about 16% is possible in a doping range corresponding to 0·2 Ω cm (Graff and Fischer 1979). Figure 17 shows the development of efficiency over time. An increase is demonstrated over one decade from about 10 to 15%, which can currently be demonstrated in the laboratory. Further expected gains cannot be dramatic, but will only be within the optimisation of technological compromises which are incorporated into present cell design.

4. Terrestrial solar cells

To make photovoltaic solar energy conversion economically attractive, the cost of solar cells must be reduced by more than one order of magnitude. The present technology of fabricating Si solar cells is based on using single-crystal Si wafers, where the single crystal is pulled out of highly purified semiconductor-grade polysilicon. In 1979, the price for these solar cells of 1 W equivalent electrical power was about 40 DM/W. Nearly 50% of this cost is contributed by the single-crystal slice. So to reduce the total cost substantially a dramatic reduction in the cost of the material is required. Numerous efforts have been started to achieve this goal by using Si with a higher concentration of impurities (so-called solar-grade Si) and sheets with a higher degree of crystallographic defects up to polycrystalline materials. Various alternative processes which are currently pursued for producing large-area Si sheet material suitable for low-cost, high-efficiency solar cells are described below.

4.1. Shaped ribbon technology

Capillary die growth, or the edge-defined film-fed growth (EFG) technique is based on feeding molten Si through a slotted die, as illustrated in figure 18. With this method, the shape of the ribbon is determined by the contact of molten Si with the other edge of the die. The die material, typically graphite, has to be wetted by the melt to result in a capillary action. Major problems are thermal instability of the process, which increases with width and growth rate of the ribbon, twin formation and segregation of SiC particles. Extremely high stresses are incorporated in the ribbon.

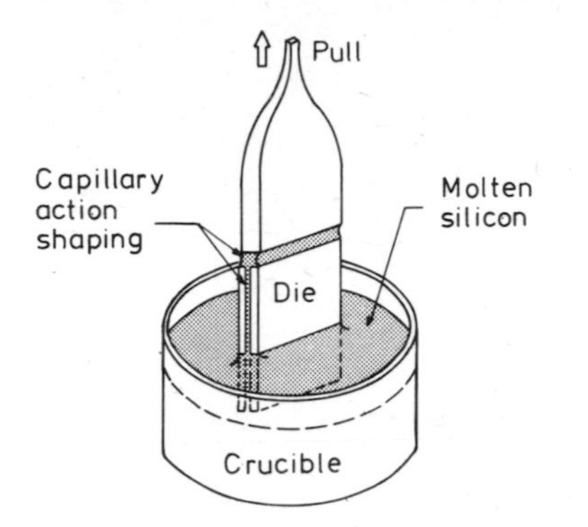

Figure 18. Capillary die growth (EFG process). Width 5 cm; growth rate (linear) 5 cm min^{-1}; growth rate (area) 1500 $cm^2 h^{-1}$.

4.2. Laser zone crystallisation process

The ribbon to ribbon process (RTR) is basically a crucible-free, float-zone growth method in which the feedstock is a polycrystalline ribbon (figure 19). The feedstock is either deposited by CVD or cast. The liquid Si is held in place by its own surface tension, and the shape of the resulting crystal is defined by the shape of the feedstock. Problems are mechanical and thermal instability, and a tendency to polycrystalline and dendritic growth.

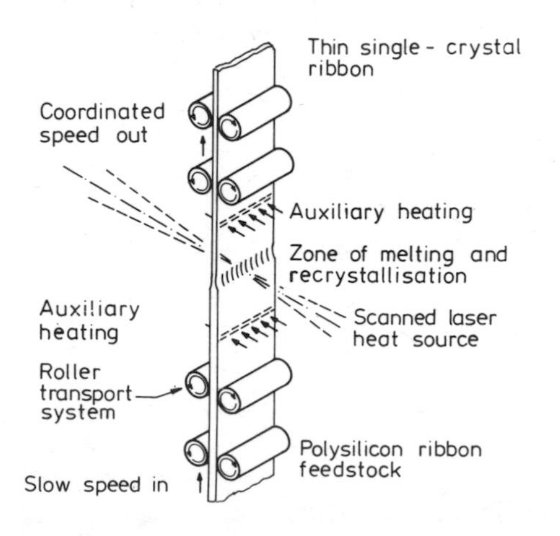

Figure 19. Laser zone crystallisation: ribbon to ribbon growth (RTR process).

4.3. Web dendritic growth

Dentritic web is a thin, wide ribbon form of single-crystal Si (figure 20). ‘Dendritic’ refers to the two wire-like dendrites on either side of the ribbon, and ‘web’ refers to the Si

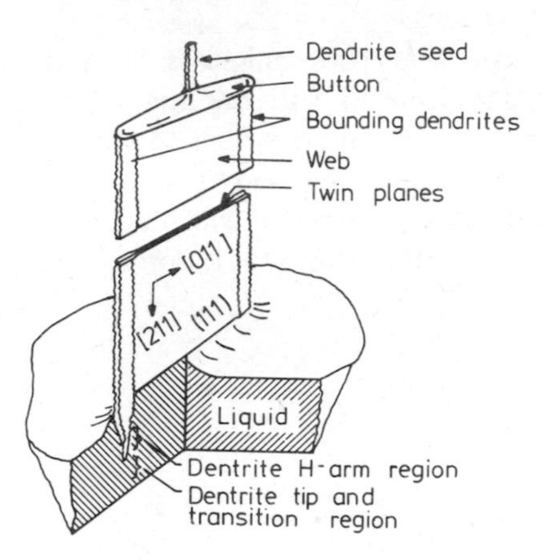

Figure 20. Web dendritic growth. Width 3·5 cm; growth rate (linear) 2·5 cm min^{-1}; growth rate (area) 500 cm^2 h^{-1}.

sheet that results from the freezing of the liquid film supported by the bounding dendrites. The best single-crystalline films so far have been produced by this method. Problems are associated with the thermal stability of the process and very low limited growth rate.

4.4. Casting of non-single-crystalline Si

At present the most advanced method of creating low-cost Si is the casting technique (figure 21). This process allows fabrication at a large volume per unit time, at least 20 times better than with conventional pulling methods (Authier 1978). The cast blocks are cut into square slices; the direct casting of Si foils will be considered later.

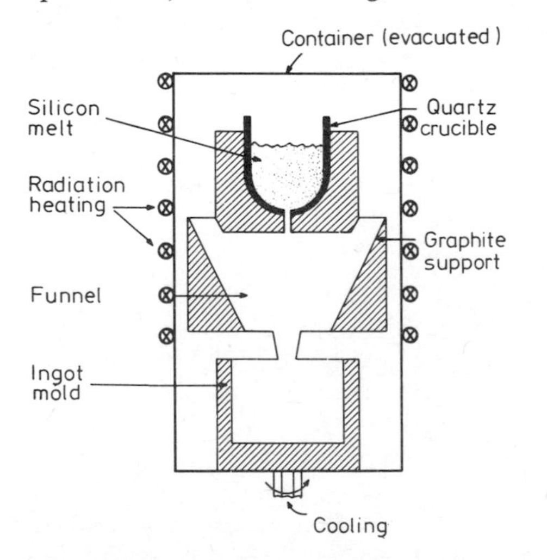

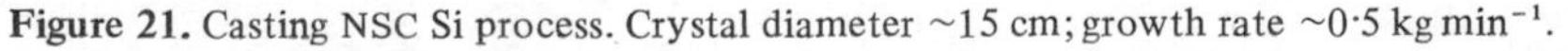

Figure 21. Casting NSC Si process. Crystal diameter ~15 cm; growth rate ~0·5 kg min^{-1}.

4.5. Si and energy yield

The future cost potentials of the two casting methods result from their possible high yields of Si material and low energy consumption compared with conventional single-crystal pulling methods. Figure 22 illustrates that situation for the cast Si, where the Si

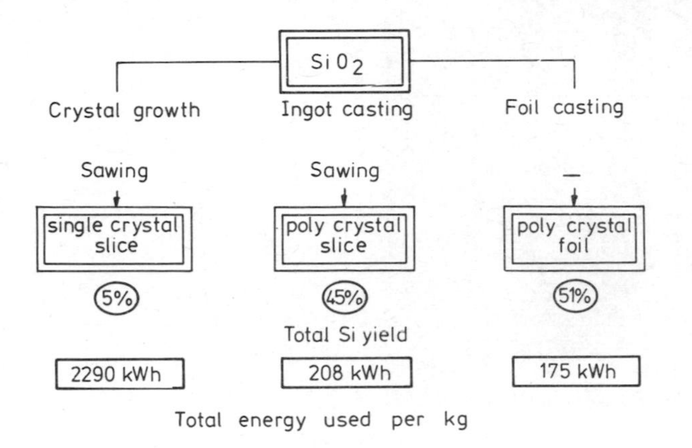

Figure 22. Si yield and total energy consumption for single-crystal growth, ingot casting and foil casting.

yield is 45% for the cast and wafered ingot, and exceeds 50% for the directly cast foil. This is in contrast to only 5% for conventionally grown Si. The energy required for 1 kg of slice material decreases from more than 2000 kWh to below 200 kWh for the new processes.

4.6. Energy pay-back time

The energy pay-back time is that time required for a solar cell to convert that amount of energy required for their production. 1 m^2 of solar cells with 10% conversion efficiency converts about 175 kWh per year in Europe. 1 kg of Si slices results typically in 1·5 m^2 of solar cells. Using the energy values shown in figure 22 for their production, the result is an energy pay-back time for a 12% single-crystalline cell which exceeds eight years, whereas for a 10% non-single-crystalline cell this value is about 14 months, and below 12 months for the foil cell.

4.7. Low-cost solar cells

The decisive question as to whether low-grade Si can be used for solar cells depends on the achievement of a solar cell exhibiting more than 10% conversion efficiency. This question has successfully been answered for the cast non-single-crystalline Si, which is characterised by controlled size and structure of the individual grains and a solar cell preparation technique which is optimised in respect of the unique characteristics of this material (Fischer and Pschunder 1977). If the individual grains of a cast wafer are columnar-oriented, with the grain boundaries mainly perpendicular to the surface, the total cell can be thought of as a parallel combination of small, individual cells. These small cells act in a normal manner, except that photogenerated carriers recombine at the side of the column as well as on the front surface and the rear side contact. Figure 24 displays a simple model of an isolated grain. A 'three-dimensional' analysis of photogenerated

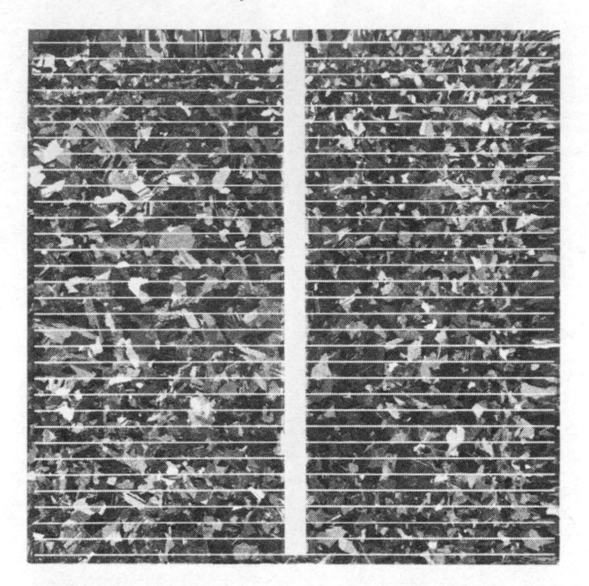

Figure 23. 10 × 10 cm² Si solar cell with an efficiency of about 10% produced from ingot casting material.

carrier collection does not exist at the moment, but in a simple one-dimensional approach an effective diffusion length can be defined. This parameter depends mainly on grain size, but also on grain boundary recombination velocity. Figure 25 shows AM1 efficiency as a function of grain size for low and high grain boundary recombination. Note that if the grain size becomes greater than bulk diffusion length and cell thickness, efficiency values exceeding 10% can be achieved. The dependence on grain size can be reduced if grain boundary recombination can be reduced. Figure 26 displays a possible scheme, if grain boundaries are p^+ doped in a p-material, minorities are reflected from the grain boundary instead of recombining there, and thus the overall grain boundary recombination velocity is reduced.

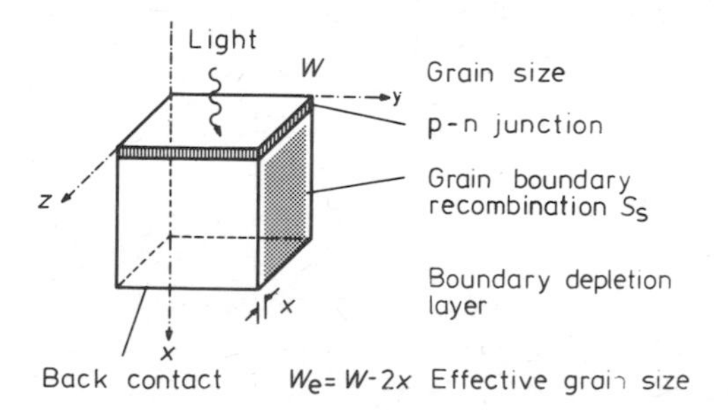

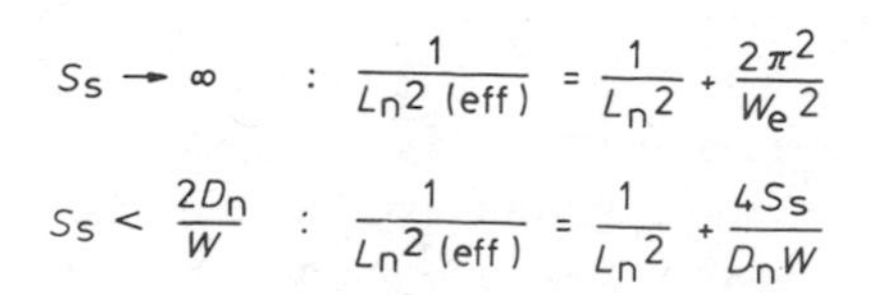

Figure 24. One-dimensional model for the influence of grain size on the effective diffusion length, L_n(eff).

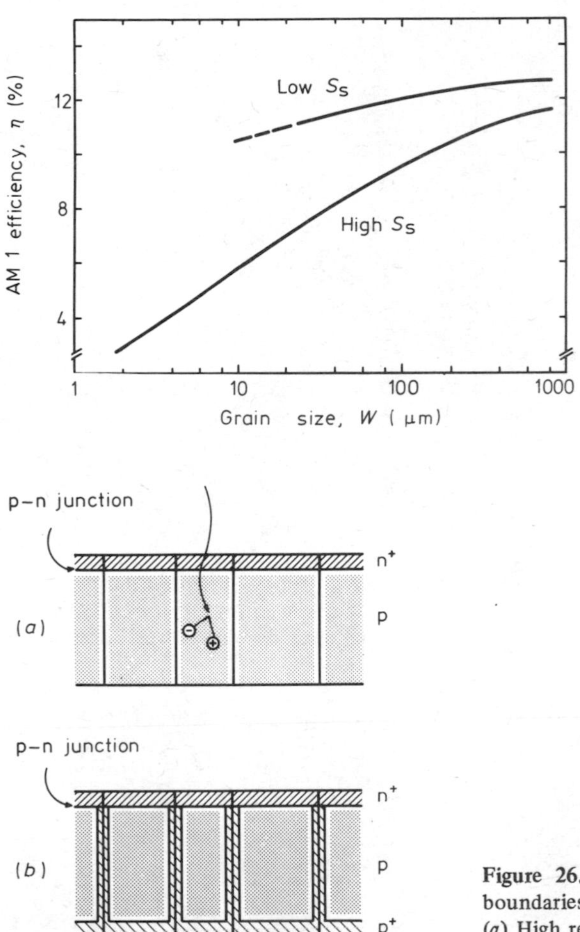

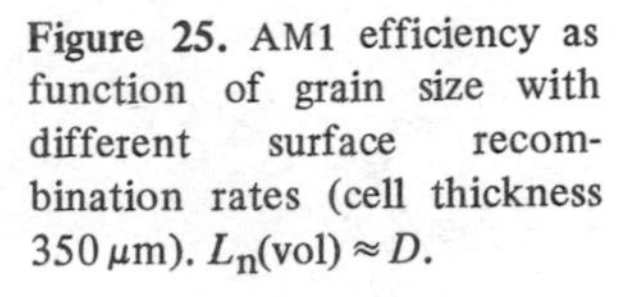
Figure 25. AM1 efficiency as function of grain size with different surface recombination rates (cell thickness 350 μm). L_n(vol) ≈ D.

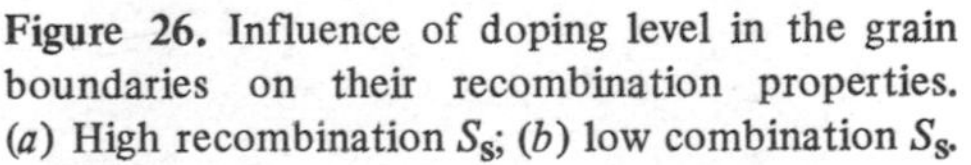
Figure 26. Influence of doping level in the grain boundaries on their recombination properties. (*a*) High recombination S_s; (*b*) low combination S_s.

4.8. Non-single-crystalline cell technology

The investigations currently under way concentrate on the optimisation of the casting process to create large ingots with a high yield of columnar-oriented volume zones and the optimisation of the solar cell process to minimise the influence of grain boundaries. Figure 27 displays the crystalline structure of different crystalline zones appearing in a cast ingot as a result of various local temperatures and freezing conditions.

Columnar-oriented structures give the desired optimum material. Cells from the transcrystallisation zone where the grains are parallel to the cell surface are efficient as long as crystal size is larger than cell thickness. Microcrystalline structure appears where crystallisation is delayed by segregation of impurities dissolved in the melt, e.g. carbon. The appearance of this structure even in small volumes in a cell is deleterious for the cell efficiency.

Grain boundaries are not always regions of high recombination rates, as shown in figure 28. The non-single-crystalline surface of a solar cell is shown in figure 28(*b*); dark areas denote a low EBIC current, and thus regions of high recombination rates. Obviously

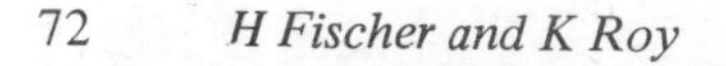

(a)

(b)

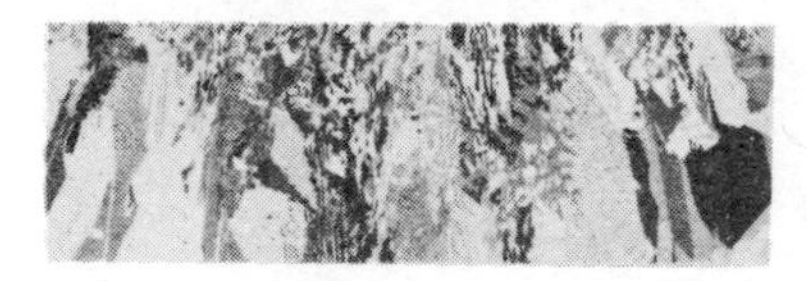

(c)

Figure 27. Different crystal structures locally distributed in cast silicon ingot: (*a*) columnar-oriented; (*b*) transcrystallisation; and (*c*) microcrystalline.

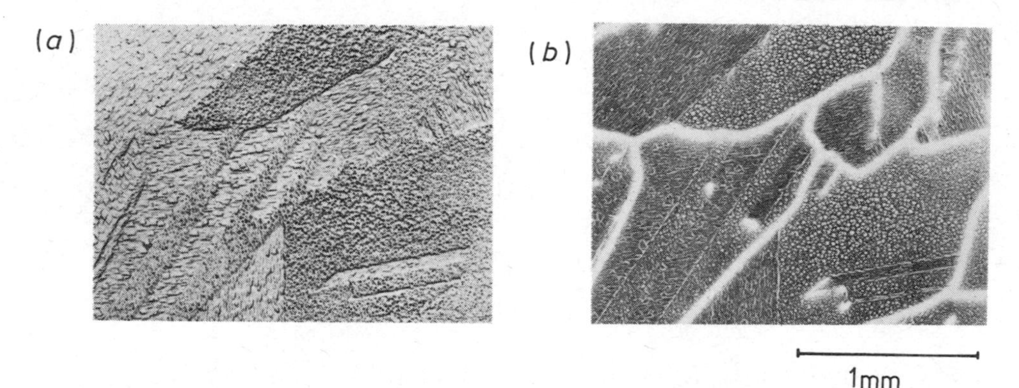

Figure 28. (*a*) SEM and (*b*) EBIC pictures of a non-single-crystalline solar cell.

the magnitude of the recombination rate depends on the degree of the crystalline disorder in the grain boundaries.

4.9. Volume production techniques

Solar cells from low-cost Si must not only be highly efficient, but must also be able to be fabricated with a high efficiency in large-volume production with extremely good yields. As a consequence, the development of economical and automated high-volume production techniques is required. Classical semiconductor fabrication methods, such as vacuum metallisation and diffusion are further developed to high throughput processes, with batch sizes up to several m^2 (that means several hundred W per batch). Continuous processes such as screen printing of metal contacts or ion implan-

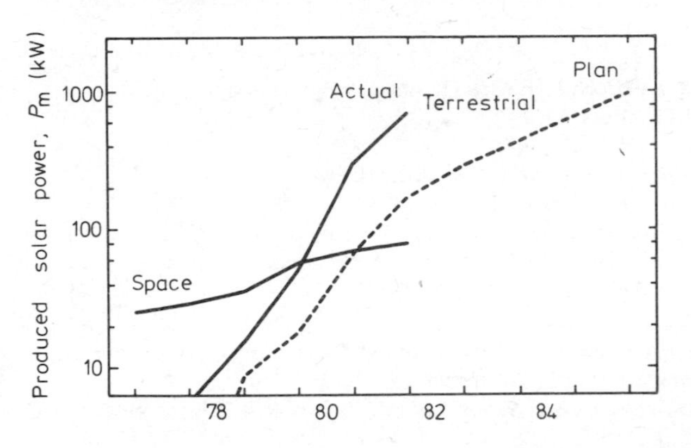

Figure 29. Volume of produced and expected prototype solar cells in accumulated electrical power from 1978 to 1985.

tation for junction formation are considered. The various fabrication methods are compared and tested in a pilot production of prototypes of terrestrial solar cells. Figure 29 shows volume of solar cells produced and expected prototypes in terms of accumulated electrical power. The volume of the current research programme is characterised by the fact that at the end of 1979 the two-year pilot line fabrication has produced an amount of solar cells equivalent to 10 years of space cell operation. The success of the programme is demonstrated by the increase in production which is expected to occur much earlier than had been previously planned. So in 1980 the same amount of Si will be used for terrestrial solar cells as for the whole range of other semiconductor products.

Prototype solar cells will be used for the production of solar generators. These generators will be tested in a wide range of terrestrial applications to investigate and demonstrate the possibility of economic photovoltaic solar energy conversion.

5. Conclusions

This paper summarises the state of the art and current development trends of solar cell technology. Among the various approaches discussed, only the Si solar cell has reached a level where it can be considered as a product and a practical semiconductor device for energy conversion in space and for terrestrial applications. Before the arrival of the energy crisis, solar cell technology was only a matter for some experts, but the field is now expanding rapidly. With the successful demonstration of high-efficiency cells from low-grade materials, a tremendous boom in the commercial development of this process has been started. From that, it can be expected that the solar cell will be a viable component in future energy concepts.

Acknowledgments

This paper owes its existence to many contributions of the solar cell group of AEG-Telefunken. The outstanding contributions of the old-timers like W Pschunder are also acknowledged, as well as those of the newcomers to the field like Dr K D Rasch and his engineering staff, and the materials people working with Dr K Graff. The permanent financial support of the 'Bundesministerium für Forschung und Technologie' is also gratefully acknowledged.

References

Authier B 1978 in *Festkörperprobleme* XVIII ed J Treusch (Braunschweig: Vieweg)
Beck J D and Conrad R 1973 *Solid St. Commun.* **13** 93
Bucher E 1978 *Appl. Phys.* 17 1–25
Conference Recs. 1976, 1978 *Photovoltaic Spec. Conf.* (New York: IEEE)
Dunbar J R and Hauser P M 1977 *IEEE Trans. Electron Devices* **24** 4
Fischer H 1974 in *Festkörperprobleme* XIV ed J Queiser (Oxford: Pergamon)
Fischer H and Pschunder W 1974 *Proc. Int. Conf. Photovoltaic Power Generation, Hamburg*
——1975 *Conf. Rec. 11th Photovoltaic Spec. Conf.* (New York: IEEE)
——1977 *IEEE Trans. Electron. Devices* **24** 438–41
Graff J K and Fischer H 1979 in *Topics of Applied Physics* vol. 31 *Solar Energy Conversion, Solid State Physics Aspects* ed B O Seraphin (Berlin: Springer)
Hovel H J 1975 *Solar Cells* in *Semiconductors and Semimetals* vol. 11 ed R K Willardson and A C Beer (New York: Academic Press)
Lindmayer J and Allison J F 1973 *COMSAT Tech. Rep.* **3**, 1
Mandelkorn J and Lamneck J-H 1972 *Conf. Rec. 9th Photovoltaic Spec. Conf.* (New York: IEEE)
Rasch K D, Roy K and Tentscher K H 1979 *Proc. 2nd E C Photovoltaic Solar Energy Conference, Berlin* pp205–12

Prospects of gigabit logic for GaAs FETs

R Zuleeg

McDonnell Douglas Astronautics Company, Huntington Beach, CA 92647, USA

Abstract. The inherent properties of GaAs FET technology are assessed which qualify its application to gigabit logic for large scale integration. Although the processing technology is not advanced enough to produce the described optimum logic gate configuration, it is demonstrated that 1 μm channel length FETs produce logic gates for gigabit logic performance with propagation delay-power products of less than 100 fJ, a prerequisite for large scale integration. Several exploratory designs will be discussed which have achieved gigabit data rate handling capability not accessible with Si device technology.

1. Introduction

Semiconductor technology is striving to achieve performance at gigabit rates in logic operations. This goal requires microwave devices in monolithically integrated circuits. As device dimensions become smaller for faster operation, the requirement for low power dissipation per device, to prevent excessive heating of the chip, becomes increasingly stringent if large scale integration (LSI) is to be accomplished. The 1980s promise to be another era for semiconductors and especially integrated circuits with VLSI as the goal. The prospects for GaAs FETs to secure a position in this development are good and success depends entirely on the availability of a high yield technology for large device density circuit fabrication.

With gigabit electronics (Bosch 1979) emerging as a branch of semiconductor technology, GaAs is certainly gaining momentum over Si when focusing on the prerequisites to satisfy the requirements for LSI. Low power and high speed small scale integrated (SSI) and medium scale integrated (MSI) circuits have been fabricated with depletion mode Schottky barrier FETs (Liechti 1977, Eden and Welch 1977), with enhancement mode junction FETs (Zuleeg *et al* 1978), and enhancement mode Schottky barrier FETs (Fukuta *et al* 1977). A planar GaAs integrated circuit technology for 1 μm channel geometries was realised for the enhancement mode JFET (Zuleeg *et al* 1978) and the depletion mode MESFET (Eden *et al* 1978) by means of selective ion implantation of n- and p-type impurities into semi-insulating GaAs substrate materials. Figure 1 presents a SEM picture of a portion of a planar GaAs positive edge triggered flip-flop fabricated with 2·5 and 1 μm channel E-JFETs. The ion implantation process is the key to low threshold voltages for E-JFETs and low pinch-off voltages for D-MESFETs in order to achieve low power dissipation per gate and control of device uniformity over large wafer sizes. Photolithography imposes limits on fabrication yields, level of integration for the 1 μm channel devices and small linewidth interconnections. Electron-beam lithography promises to overcome these constraints and will open new perspectives for LSI and eventually VLSI.

0305-2346/80/0053-0075 $01.00

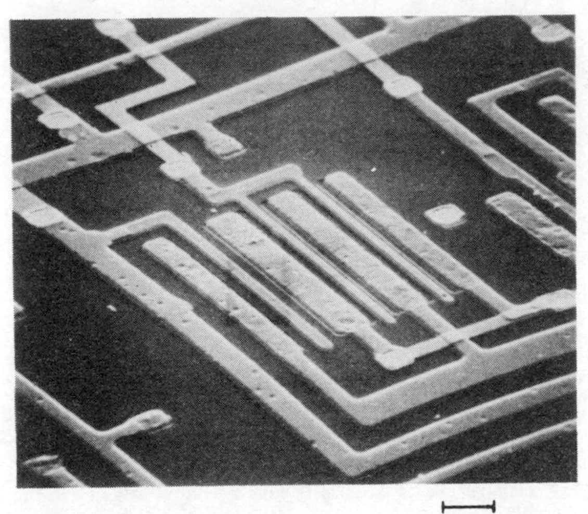

Figure 1. SEM photograph of planar IC technology with 2·5 and 1·0 μm channel E-JFETs (magnification ×1000).

This paper proposes and describes an optimum gate design for GaAs FETs utilising a two-terminal current limiter. It presents the various exploratory approaches for gate designs now under development. Projections for the 1 μm channel logic gate performances are presented and extrapolated to performances with 0·5 μm channel device technology.

2. Optimum gate design

In analogy to Si MOS-transistor integration practices the combination of enhancement mode FET driver with depletion mode FET load could be accomplished if appropriate processing is available to control the desirable low threshold and pinch-off voltages (figure 2*b*). With enhancement MESFET and E-JFET the resistive load is being used with existing technology (figure 2*a*) and impressive performances were obtained (Mizutani *et al* 1979). A tunnel diode was also proposed as a load (Lehovec 1979) and is under

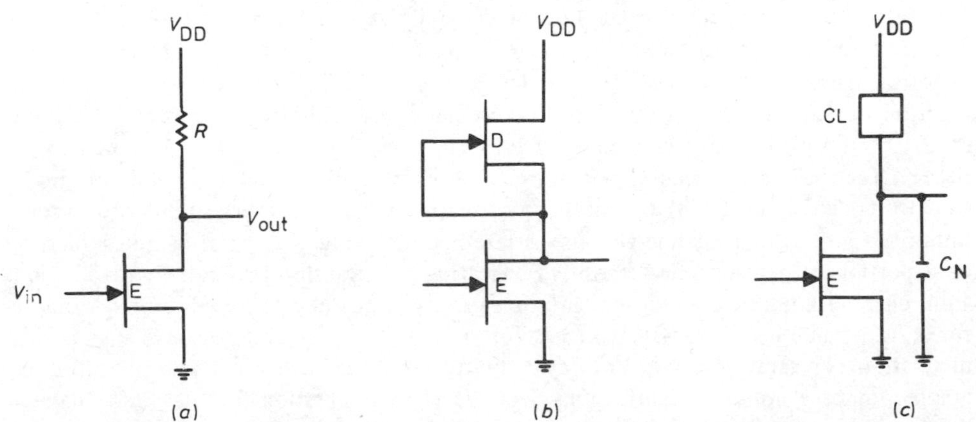

Figure 2. GaAs logic gate circuits for DCFL: (*a*) resistive load; (*b*) depletion mode load; and (*c*) current limiter load.

investigation. Although the resistive load does not offer the best switching characteristics, (Zuleeg *et al* 1978) because of asymmetrical rise- and fall-time, the 'resistor' changes advantageously into a high speed current limiter through proper geometrical design and the inherent material properties of GaAs (figure 2*c*). The current under drift velocity saturation of this constant current source is then

$$I_L = qNav_{lim}W \tag{1}$$

and the voltage for saturation is approximately

$$V_S \simeq E_M L + R_c I_L \tag{2}$$

where W is the width, a the height, N the doping concentration, v_{lim} the scattering-limited drift velocity, E_M the critical electric field and R_c is the contact resistance. Figure 3 gives the geometry and experimental voltage–current characteristics of n^+nn^+ devices with $W = 2{\cdot}5$ and $5{\cdot}0\,\mu m$ and $L \simeq 1\,\mu m$ using an ion implantation dose of Si^+ equal to $Na = 2 \times 10^{12}$ atoms/cm^2. The limiting current value is well controlled by the width W of the experimental devices, but V_S is out of range for low voltage circuits. A value of

(*a*)

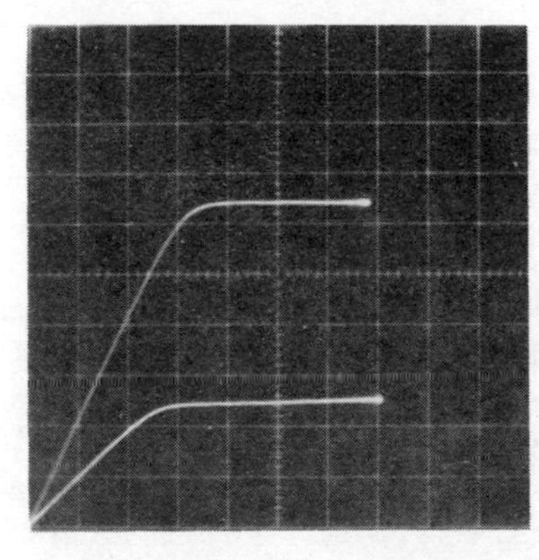

Vertical scale: 200 μa / division
Horizontal scale: 0·5 V/division

(*b*)

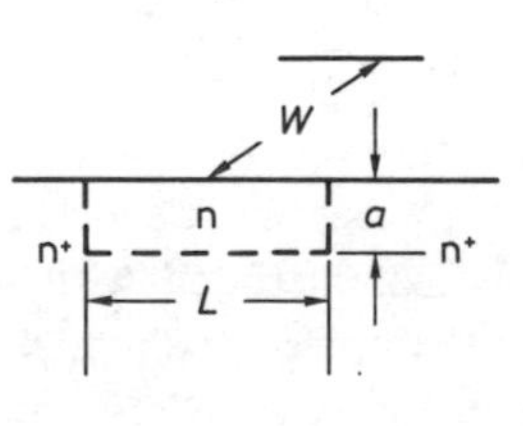

Figure 3. (*a*) Voltage–current characteristic; (*b*) geometry of GaAs high speed current limiter (see text for definitions of symbols).

Table 1. Switching performance of EJFET driver with current limiter load element ($P_D = 200\,\mu W$; $W = 12\,\mu m$; $V_{TH} = 0{\cdot}25$ V).

FET driver L (μm)	FO	t_{pd}(ps)	t_R(ps)	t_F(ps)	$P_D \times t_{pd}$(fJ)	$f_{CL} = \frac{1}{t_{pd} + t_R}$ (GHz)
1·0	1	55	80	140	11	7·4
	3	118	190	280	23·6	3·4
0·5	1	35	60	100	7	10·6
	3	60	100	160	12	6.3

0·5 V or less is desirable for low threshold and supply voltage circuits to make this element effective. Computer simulated performance characteristics of an optimised logic gate using the current limiter load with an E-JFET driver for a power dissipation of 200 μW with $V_{DD} = 1{\cdot}5$ V are presented in table 1. These numbers in table 1 reflect the fan-out (FO) capability and the effect of channel length reduction to 0·5 μm on speed performance. Since for ultimate switching performance of logic gates for LSI one requires a low node capacitance C_N, a large current drive capability, a small device size and low component count per gate, this optimum gate design in GaAs has superiority over any other Si and GaAs design. Because the value of K' is greater by a factor of 5–10 in the voltage–current relation of a GaAs FET (Zuleeg *et al* 1978, Eden *et al* 1979),

$$I_D = K'\left(\frac{W}{L}\right)(V_G - V_T)^2 \quad , \tag{3}$$

a low switching energy for GaAs gates can be achieved for speeds not equalled with Si gates. This performance is primarily related to the larger electron mobility in GaAs. The current limiter offers the lowest node capacitance of any logic gate design for GaAs. Although not yet practical with existing technology, this 'ideal gate' is feasible with advances in GaAs device fabrication technology.

3. Exploratory gate designs

Until the described 'ideal gate' becomes practical, other approaches are pursued which serve the purpose of advancing technology and establishing confidence in the GaAs approach to LSI. With experimental gate structures the gigabit logic capability of GaAs FETs was demonstrated with SSI and MSI circuits. LSI is, however, only achievable by reducing the power dissipation per gate to values in the range of 100–200 μW. The logic family, exploring GaAs MSI and LSI, consists of the following members: buffered FET logic (BFL) (Liechti 1977); Schottky diode FET logic (SDFL) (Eden and Welch 1977); and direct coupled FET logic (DCFL) (Zuleeg *et al* 1978, Fukuta *et al* 1977, Mizutani *et al* 1979, Notthoff and Vogelsang 1979). Figure 4 presents the basic logic gate circuits. The fastest circuits today are designed with BFL (Liechti 1977) and operate at clockrates as high as 4·5 GHz with rather high power dissipations of 10–20 mW/gate which prohibit LSI. With design optimisation this approach may, however, yield MSI logic above 4 Gbit/s with reduced power dissipation. Proponents of the SDFL approach are claiming 1–2 Gbit/s MSI capability (Eden *et al* 1979) with a power dissipation in the range 0·2–2 mW/gate. An integration complexity of 64 gates on a chip was achieved and LSI is projected as the next milestone. The DCFL approach is pursued using the E-MESFET (Fukuta *et al* 1977, Mizutani 1979) and E-JFET (Zuleeg *et al* 1979, Notthoff and Vogelsang 1979) and so far has only reached SSI complexity status with MSI under development. DCFL can claim to possess the prerequisites for high speed LSI, offers advantages in fabrication integration by direct coupling without level shifting and has low power dissipation. E-MESFETs with resistive loads give delay-power products of 75 fJ (Mizutani 1979) and DCFL designed with E-JFETs in the follower–inverter configuration operates with 1–2 Gbit/s data rates and delay-power products of 50 fJ (Notthoff and Vogelsang 1979). MSI circuits are under development in both technologies for DCFL. The E-JFET logic has a considerable edge over the E-MESFET counterpart in noise immunity.

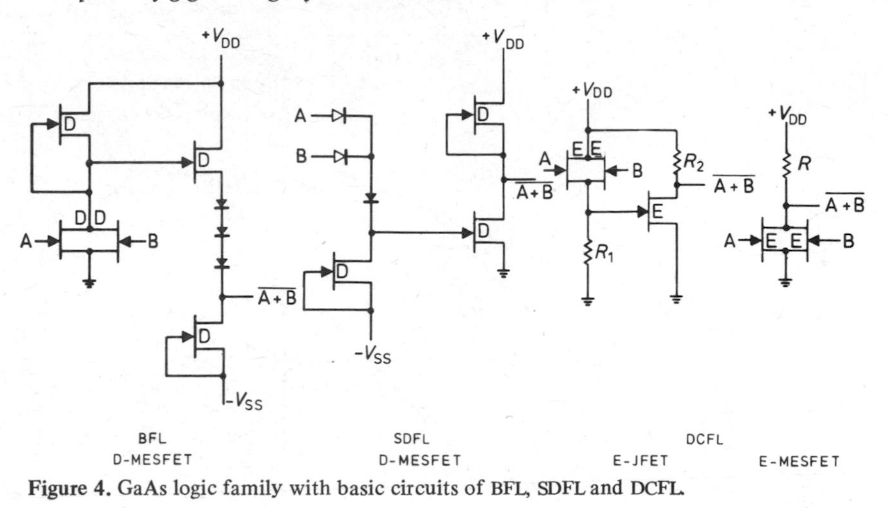

Figure 4. GaAs logic family with basic circuits of BFL, SDFL and DCFL.

4. Discussion and conclusion

Projections for the GaAs logic family are presented in figure 5, which also contrasts the GaAs performance with the Si technology based integrated circuits. To qualify for gigabit LSI a propagation delay-power product of less than 100 fJ is required, hence the delay-power product per gate is the important figure of merit for characterising the logic gate performances. The dominance in speed of GaAs is obvious from figure 5 and combined with low power dissipation yields a trade-off between Si and GaAs. With a choice of equal power dissipation, GaAs circuits are 4–5 times faster than Si ones and for equal high switching speeds a vastly reduced power consumption for GaAs circuits is possible, thus establishing a basis for GaAs gigabit logic at LSI levels. Electron-beam technology for

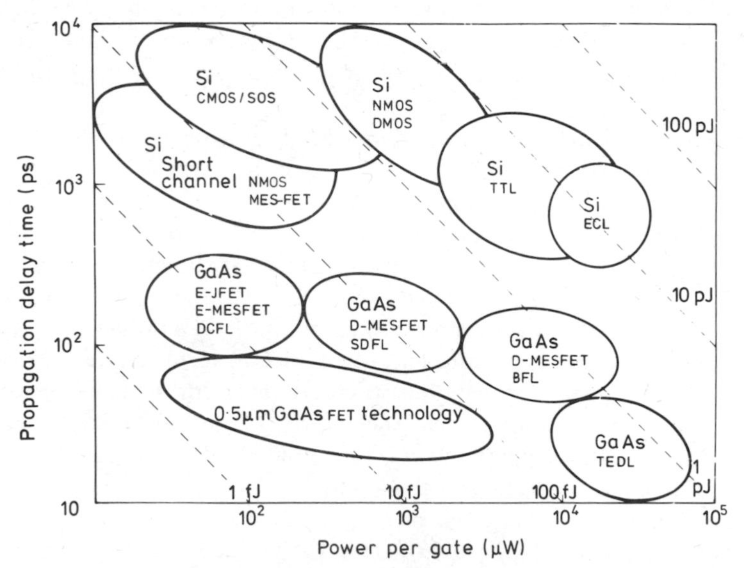

Figure 5. Propagation delay time against power dissipation for GaAs IC technologies in comparison to Si devices and IC technology.

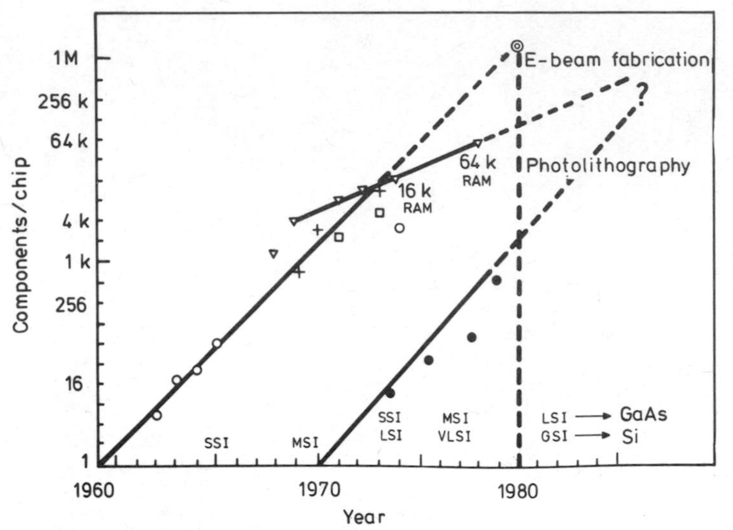

Figure 6. Components per chip advances over the years for Si and GaAs integrated circuit technology. ● GaAs FET logic; ▿ MOS arrays; □ MOS logic; ○ bipolar logic; + bipolar arrays.

0·5 μm FETs is projected to produce LSI with improved speed capabilities as shown in figure 5.

Considerable efforts are deemed necessary in: (i) logic gate optimisation to obtain maximum speed at minimum power; (ii) exploration and refinement of processes, such as ion implantation, electron-beam lithography; and (iii) high resistivity substrate material growth and perhaps molecular beam epitaxy, before GaAs technology reaches the maturity of Si technology now successfully applied to produce VLSI. Just as the planar technology has triggered the evolution of Si integration technology, now highlighted by VLSI, the planar GaAs technology has established the vanishing point of a perspective for GaAs integrated circuits with excellent prospects for gigabit LSI to unfold in the 1980s.

Looking at the factual data of Si integration progress in figure 6, one sees immediately that according to Moore's analysis (Moore 1975) Si technology has developed at a rate such that the number of components per chip has been doubling every year since 1960. This would extrapolate to one million for giant scale integration (GSI) by 1980. This linear logarithmic extrapolation will not continue indefinitely and design and technological limits may terminate practical hardware in the GSI region. GaAs technology parallels that of Si with a time scale transformation of 10 years. While Si is capitalising on a mature technology, GaAs is struggling to advance to maturity. The LSI capability of GaAs projected in figure 6 for the 1980s is confronted with the same three kinds of limits as Si to constrain the development. They are: (1) fundamental limits imposed by physical laws; (2) complexity limits relating to problems and economy of design; and (3) technological boundaries which gate the advancing development. It is really the third factor which is of concern to the GaAs future. The second factor applies to both technologies. Several things will act in favour of GaAs LSI when considering the limitations of Si devices in comparison to GaAs ones from the point of view of the first factor: there are scaling problems for Si devices, which are not prevalent in GaAs devices, and there are fundamental speed advantages for GaAs devices. Since gigabit logic will only be achieved in fully integrated LSI form, this leaves GaAs on the horizon to accomplish this goal. Experimental logic gates have demonstrated 2–4·5 Gbit/s logic with high power and 1–2 Gbit/s logic with medium and low power dissipation, depending on the particular technology used. A theoretical 'optimised logic gate' promises 5–10 Gbit/s logic with low power dissipation for LSI when technological advances make it practical to achieve.

References

Bosch B G 1979 *Proc. IEEE* **67** 340–79

Eden R C and Welch B M 1977 *IEEE Trans. Electron Devices* **ED24** 1209–10

Eden R C, Welch B M and Zucca R 1978 *IEEE J. Solid St. Circ.* **SC13** 419–26

Eden R C, Welch B M, Zucca R and Long S I 1979 *IEEE Trans. Electron Devices* **ED26** 299–317

Fukuta M, Nyama S and Kusakawa H 1977 *IEEE Trans. Electron Devices* **ED24** 1209–14

Lehovec K 1979 *1st Spec. Conf. on Gigabit Logic for Microwave Systems, Orlando, Florida, 1979, Tech. Abstracts* 102–6

Liechti C H 1977 *Gallium Arsenide and Related Compounds (Edinburgh) 1976* (Inst. Phys. Conf. Ser. 33a) pp227–36

Mizutani T, Ida M and Ohmori M 1979 *1st Spec. Conf. on Gigabit Logic for Microwave Systems, Orlando, Florida, 1979, Tech. Abstracts* 93–101

Moore G E 1975 *IEDM Dig.* 11–12

Notthoff J K and Vogelsang C 1979 *1st Annual IEEE GaAs IC Symp., Lake Tahoe, Nevada, 1979, Res. Abstracts paper* 10

Zuleeg R, Notthoff J K and Lehovec K 1978 *IEEE Trans. Electron Devices* **ED25** 628–39

Trends of MOS devices and circuit elements

Kurt Hoffmann

Siemens AG, Component Division, Balanstrasse 73, Munich, West Germany

Abstract. Starting in 1967 with hundreds of components per MOS circuit, the integration level rose toward today's very large scale integrated (VLSI) circuits with tens of thousands of components per chip. Two major device development trends which lead toward these circuits can be distinguished: the scaling of existing conventional devices and the indirectly generated short-channel devices such as DMOS and VMOS. In conjunction with this development the product evolution is considered by analysing the trend of basic circuit elements for digital and analog circuits.

1. Introduction

The original concept of a field effect transistor can be traced back to a US patent granted to Lilienfeld in 1930 (figure 1). From his patent, however, it is not clear how the control mechanism of the transistor was supposed to function. The MOS capacitor structure as it is known today was first described by Heil in 1935. Since that time many researchers have considered similar structures. A not very well known fact is that in 1945 Welker had already derived the current equation of the MOS transistor for the linear region (Weiss 1975), but it was not until 1967 that the major technological problems were overcome.

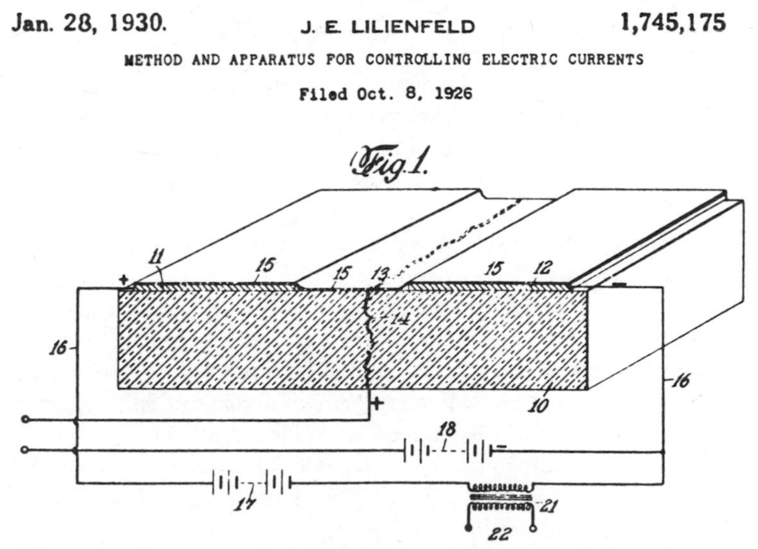

Figure 1. Schematic of Lilienfeld's MOS transistor: 10 insulating substrate; 11, 12 source, drain; 13 control electrode; 15 conductive coating.

0305-2346/80/0053-0083 $02.00

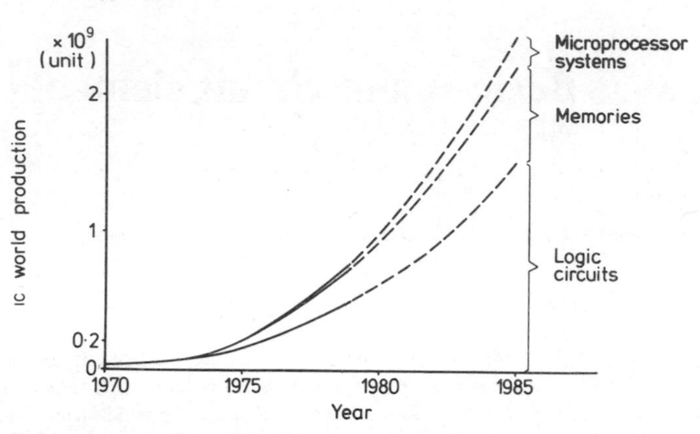

Figure 2. Annual world-wide production of MOS ICs with the contributions of various product lines.

Since then technical improvements, increase in volume production and cost reduction have mutually enhanced and led to a vigorous growth in production of MOS–ICs (figure 2) with sales estimated to be $2400 million world-wide in 1979. This amounts to about 50% of the total IC sales (Siemens AG 1979).

True appreciation of this growth rate, however, can only be achieved when the rising level of integration is also considered (Garbrecht and Stein 1976). Starting with hundreds of components per chip in 1967 the integration level rose toward today's very large scale integrated (VLSI) circuits with tens of thousands of components per chip.

In this paper the MOS devices and circuit elements which led toward this development will be analysed and discussed.

2. MOS device development

2.1. Directly generated conventional devices with small dimensions

The major technical driving force behind the production growth rate of MOS circuits is the ability to increase the integration level by scaling the MOS transistor. Scaling is a concept that concerns the coordinated changes in dimensions, voltages and doping concentrations by a factor α (Dennard *et al* 1974). A simplified summary of the scaling properties is shown in table 1.

The most significant consequence of scaling is that the component density increases by α^2, the delay time decreases by $1/\alpha$ and the power dissipation per unit area remains unchanged. The last statement means that no power dissipation problem will be encountered in the forseeable future despite the α^2 increase in component density.

The practical realisation of the scaling principle is limited by technological shortcomings associated with production techniques, circuit design difficulties and by physical parameters which cannot be scaled (Rideout 1978, Wallmark 1975).

Some of these physical parameters are the temperature (since in view of the high cost cooling is not practical), the energy gap, the Fermi potential and the depletion layer width which can only be scaled to a limited degree.

The technological limits which have to be overcome in the near future are e.g. the generation and control of small linewidth and spacing, and thin gate oxides free of

Table 1. Scaling properties.

Transformed quantities	
Lateral and vertical dimensions	$1/\alpha$
Supply voltages	$1/\alpha$
Doping concentrations	α
Results	
Component density	α^2
Power dissipation per function	$1/\alpha^2$
Power per area	1
Delay time	$1/\alpha$
Power delay product	$1/\alpha^3$

defects. In solving these problems particular emphasis has to be given to the reliability and the yield of the process. Some technical solutions such as the projection lithograph, plasma etching, multiple ion implantation and low-pressure CVD techniques are very promising.

The predominant circuit design problems are the management of the gigantic number of components on a chip (Rüchardt 1978) and the restraints associated with some electrical parameters. Some of these are the increased resistance of diffused regions, the limited current capability of metallisation lines caused by electromigration and small geometry effects of the MOS transistors. An example of one small geometry effect is the threshold voltage dependence on the device geometry (Wang 1977, 1978, Poon *et al* 1973) as shown for an n-channel transistor in figures 3 and 4.

At short-channel length L the depletion region width DW of source and drain is not negligible. The positive donors of source and drain support the positive gate charge (inset of figure 3) which causes the threshold voltage to be decreased (figure 3). In contrast to this decrease is the threshold voltage increase with narrow-width devices (figure 4). With increasing gate voltage the gate depletion region spreads into the substrate (inset figure 4).

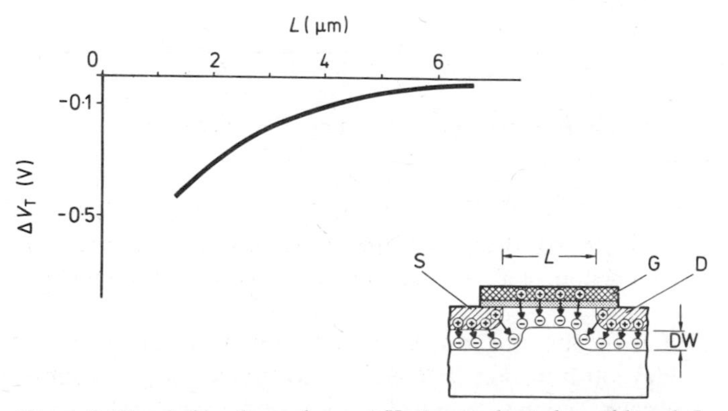

Figure 3. Threshold voltage change ΔV_T due to short-channel length L.

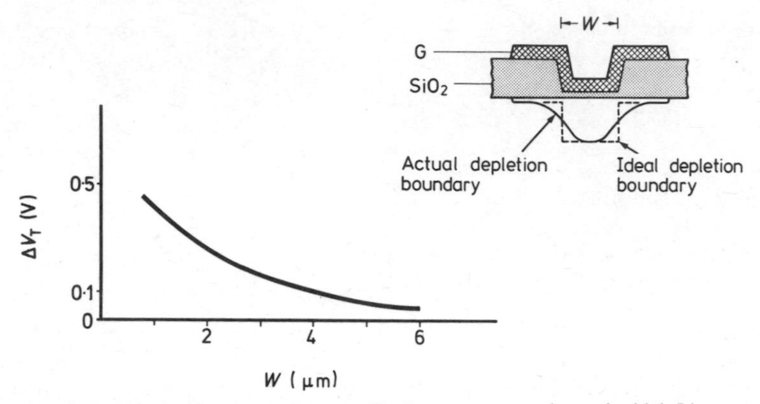

Figure 4. Threshold voltage change ΔV_T due to narrow-channel width W.

Table 2. Typical MOS process parameters in 1979.

	Production	Pilot production
Feature size	3·5–5 μm	2–3 μm
Edge position accuracy	±1·5 μm	±1·0 μm
Gate oxide	1000–1200 Å	400–600 Å
Junction depth	1·2 μm	0·6 μm
Sheet resistivity diffusion layer	15 Ω	25 Ω

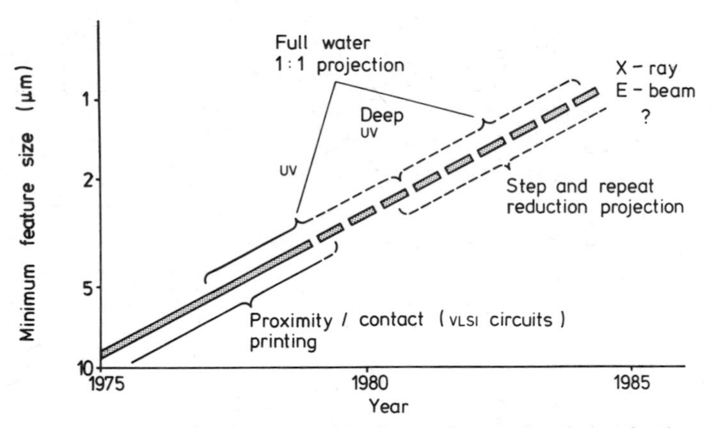

Figure 5. Past and possible future minimum feature sizes in production with various lithographic methods.

However, the depletion boundary diverges from the ideal square boundary. In narrow-width devices this phenomenon tends to prevent inversion and an increased threshold voltage results.

Despite the difficulties encountered in scaling the MOS transistor, it is remarkable what has been achieved. Today's feature size (linewidth and spacing) of 3·5–5 μm is standard for a production process and 2–3 μm are in use in pilot production (table 2).

The development of the feature size (figure 5) is of particular interest since its reduction determines the scaling rate factor for the foreseeable future until about 1985. At the present time contact and proximity printing are still in heavy use. 1 : 1 projection and step and repeat reduction projection are the most popular lithographic methods for the near future (Rideout 1978). For feature sizes below 1 μm electron beam or x-ray methods are on the horizon (Lyman 1979).

2.2. *Indirectly generated short-channel devices*

In the previous section a description was given of how scaled conventional planar MOS devices can be generated with the use of small geometries. Another development trend concerned mainly with the improvement of MOS circuit performance, like high-speed or high-voltage operation, is the indirect generation of short channel dimensions using ordinary photolithography. Many methods are being investigated. The devices best known are the double diffused MOS transistor (DMOS) (Tarui *et al* 1971) and the vertical MOS transistor (VMOS) (Rodgers *et al* 1974).

The DMOS transistor (figure 6*a*) is fabricated by the use of double diffusion through the same oxide window. The doping profile (figure 6*b*) varies along the channel, as is common for all short-channel devices indirectly generated. The p-region defines the short channel typically below 1 μm. A slightly doped π-region, the drift region, serves as a space-charge region to separate the n^+ drain from the p-region. This lowers the drain parasitic capacitances and increases the breakdown voltage of the drain–substrate junction. A series combination of two transistors with different threshold voltages can be used as a simple model (Rodgers *et al* 1975) to describe the DMOS transistor (figure 6*c*). In this model a short-channel enhancement transistor is in series with a long-channel depletion transistor with common gates.

Recently, double implantation techniques have been used to improve further the DMOS transistor by better controlling and adjusting of the doping profile (Tihanyi and Widmann 1977).

The other short-channel device mentioned, the VMOS transistor, uses the third dimension generated by an anisotropic v-groove etch (figure 7). The VMOS doping profile is similar to that of the DMOS transistor, but the v-groove provides the VMOS transistor with roughly twice the current capacity per unit area.

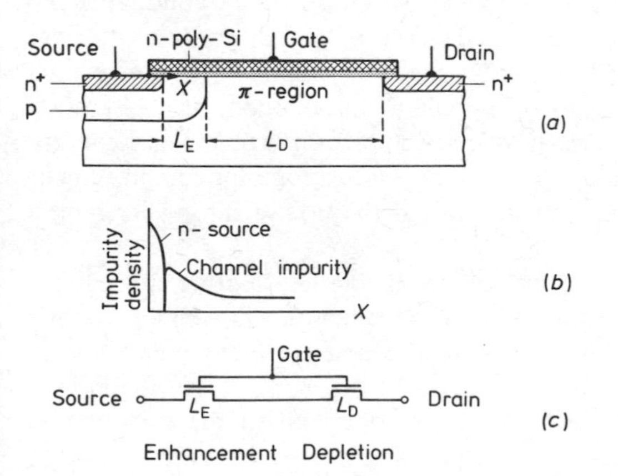

Figure 6. DMOS transistor: (*a*) cross section; (*b*) doping profile; and (*c*) two-transistor model.

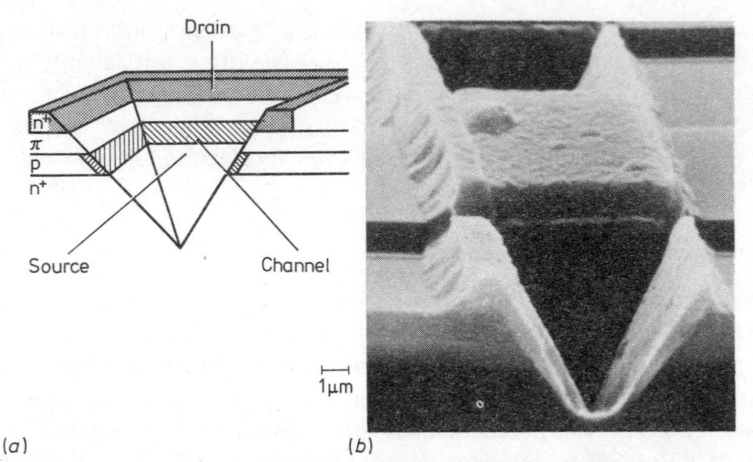

Figure 7. VMOS transistor: (*a*) schematic; and (*b*) microphotograph.

The three-dimensional feature of the VMOS transistor has particularly attracted memory designers to implement one-transistor cells. Cell sizes smaller than 150 μm^2 using 4 μm feature sizes have been reported (Hoffmann and Losehand 1978, Essel *et al* 1979).

Up to now the short-channel devices generated indirectly have been used to improve the MOS transistor performance. In the future these devices may gain in importance as some of the restraints encountered by scaling a conventional MOS transistor are more easily overcome.

3. Circuit elements

3.1. Digital circuit elements

The most fundamental digital circuit is the MOS inverter. Its switching performance is predominantly governed by a pull-up device, an enhancement or depletion transistor. As the design of such inverters is well documented there will be no further discussion here.

Next in importance for digital circuits are the currently used memory cells, such as dynamic, static and non-volatile cells. Among these cells the dynamic one-transistor cell occupies the smallest cell area, and figure 8 shows the cell implemented in double polysilicon technology. The cell area is 450 μm^2 using 5 μm feature size. One polysilicon layer is used for the address transistor TA and the other for the MOS varactor C. The cell is used in today's most sophisticated dynamic RAM in volume production, the 16k RAM (Mitterer 1979). Newer cell designs for the next generation of dynamic RAMs concentrate on the scaling of the existing structure. One promising improvement reported (Tasch *et al* 1976) is the double implantation of the MOS varactor whereby the memory capacitance is increased.

The one-transistor cell is exclusively used in dynamic memories. In other applications the cell has proved to be impractical because of the complex peripheral support circuits needed and the difficulties encountered in testing larger cell arrays in conjunction with other logic functions on a chip.

This is not so for static cells, which find use in memories, microprocessors and all kinds of logic circuits. A good part of the newer cell development is concentrated on the

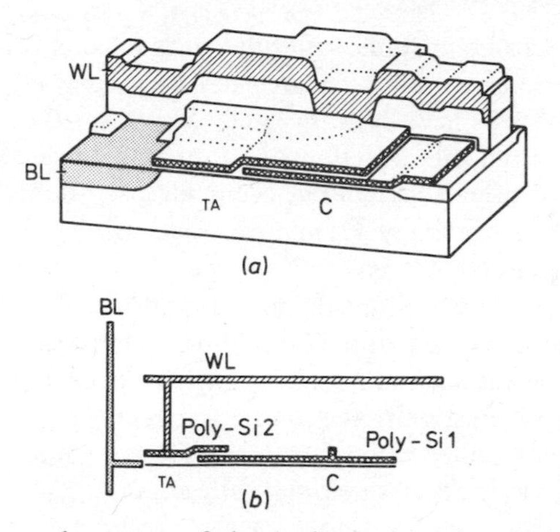

Figure 8. One-transistor cell in double polysilicon technology: (*a*) cross section; and (*b*) circuit symbolism.

replacement of the depletion-load devices by polysilicon load resistors as shown in figures 9(*a*) and (*b*). The advantages are twofold:

(i) The power dissipation of the cell can be reduced drastically by using very high resistivity polysilicon layers as load resistors. The only requirement for the resistivity value is that the 'high' level at the circuit node is maintained under worst case leakage current conditions ($I_P \gg I_L$).

(ii) The cell area as demonstrated by the layouts (figures 9*a* and *b*) can be reduced. Further progress in area reduction can be achieved by the use of a double polysilicon process, whereby the additional polysilicon layer is used as load resistor and placed on top of the pull-down transistors T_P. Cell sizes of 1000 μm^2 using 4 μm feature sizes are known (Pashley 1979).

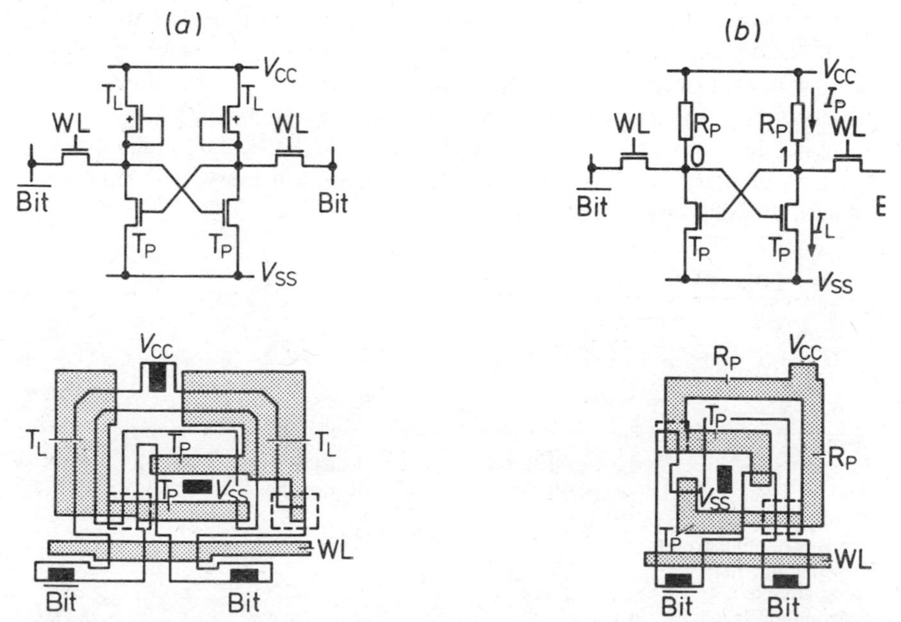

Figure 9. Schematic and layout of static memory cells: (*a*) with depletion loads; and (*b*) with polysilicon loads.

The implementation of non-volatile RAM cells compatible with other MOS circuits is still the industry's most sought after circuit element. Much research has been concentrated on this effort and many novel memory cells have been invented. At present the metal nitride oxide silicon (MNOS) and the stacked gate injection MOS (SIMOS) (Rössler and Müller 1975) memory elements (figure 10) offer a limited but economical solution. These memory elements retain the information for many years and can occasionally be reprogrammed. Their distinct features are shown in table 3.

In the SIMOS memory element programming is achieved by the injection of hot electrons from the channel into the floating gate, resulting in a large shift in threshold voltage. Erasing can be performed either by optical or electrical means, where charge is emitted from the floating gate by Fowler–Nordheim field emission to the source or drain region.

In the MNOS memory element charge carriers tunnel through a thin oxide layer into traps at the oxide–nitride interface, causing a shift in threshold voltage. Erasing is performed by electrical pulses of opposite polarity.

The MNOS memory element is used mainly in circuits for entertainment applications, such as TV-tuners, because reprogramming can be performed relatively quickly and the limited number of read cycles is no drawback.

The SIMOS memory element with its unlimited number of read cycles is particularly suitable for modern microprocessors. The cell is implemented in the widely used double polysilicon process. For the two SIMOS cells shown in table 3 the optically erasable version offers the most economical solution for high-density electrically programmable ROMS (EPROMS), whereas the electrically erasable cell also allows the reprogramming of the memory contents in the system. An example using such a cell is the 8 kbit electrically erasable and programmable ROM (EEPROM) shown with its most important features in figure 11.

The MNOS and SIMOS memory cells which are used in today's memory products have quite distinct and different electrical performances (table 3). Newer developments close this performance gap. In one development the read cycle limitation which accompanied all MNOS memory cells has been overcome with an approach whereby the MNOS gates are grounded during the read mode (Hagiwara *et al* 1979), and in another development the long programming and erase time of the SIMOS memory cells has been drastically reduced by using very thin gate oxides (West 1979).

Table 3. Electrical features of non-volatile memory cells.

	MNOS	SIMOS	SIMOS
Programming	electrically	electrically	electrically
Erasing	electrically	electrically	optically
Read cycles	10^{11}	∞	∞
Reprogramming cycles	10^7	10^4	–
Power dissipation (programming)	μW	mW	mW
Programming time	10 ms	50 ms	50 ms
Erase time	50 ms	1 s	30 min (UV)

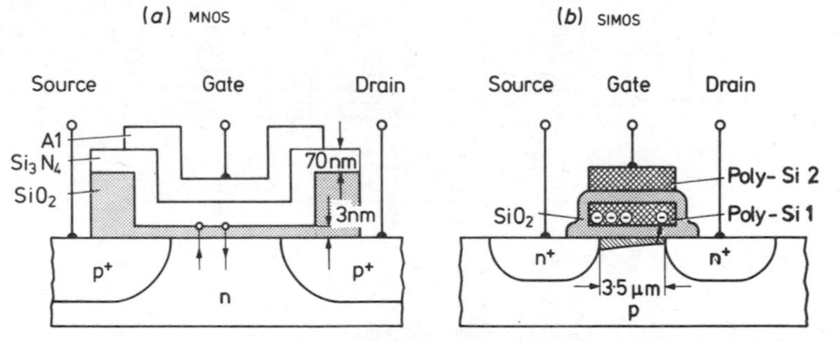

Figure 10. Cross section of non-volatile memory cells: (*a*) MNOS cell; and (*b*) SIMOS cell.

Figure 11. 8k EEPROM Siemens HYB 2808. Features: Data retention 10 years; global electrical erasure; unlimited number of read cycles; access time 450 ns.

3.2. Analog circuit elements

Until recently, very little had been done to use MOS technologies for analog circuit applications, as bipolar technology dominated this field. One reason for considering MOS analog circuitry arose with the development of MOS, LSI and VLSI circuits, which increasingly require interfacing with analog data and their processing. The key MOS analog circuit elements are charge transfer devices (CTD), precision-ratioed MOS capacitors, analog switches and operational amplifiers.

With CTDs very sophisticated analog signal processing can be performed. They can, e.g., be used as simple clocked analog delay lines or for the realisation of frequency selective filters and matched filters. As an example the cross section and implementation of a four-phase charge-coupled device (CCD) in double polysilicon technology are shown in figure 12. The electrode width was optimised to 50 μm in order to place the CCD into a small Si area and achieve a high transfer efficiency.

Precision-ratioed MOS capacitors are gaining in importance in circuit designs (McCreary and Gray 1975). They are the most precise passive circuit element available in MOS

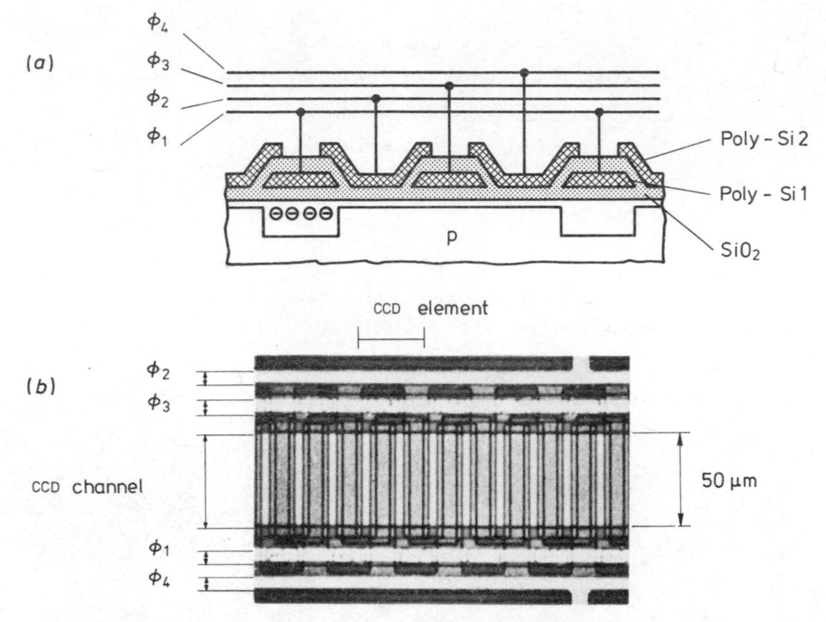

Figure 12. Charge-coupled device in double polysilicon technology: (*a*) schematic; and (*b*) realisation.

Table 4. Characteristics of integrated resistors and capacitors.

Component	Fabrication	Matching (%)	Temperature coefficient (ppm/°C)	Voltage coefficient (ppm/V)
Resistor	Diffused $W = 50\,\mu m$	±0·4	+2000	~200
	Ion-implanted $W = 40\,\mu m$	±0·12	+400	~800
Capacitor	n–Al–gate MOS $T_{OX} = 1000$ Å	±0·06	+26	~10

technology. Published data (Suárez *et al* 1975) of measured resistor and capacitor values are compared in table 4. The capacitor was implemented between a metal and heavily doped Si layer, as it is possible in an Al-gate process. In the double polysilicon process capacitors can be implemented between two layers of heavily doped Si as shown in figure 13 for a capacitor array. This approach has the advantage that parasitic capacitances are much smaller than in the first case. In either case the thermally grown oxide serves as a dielectric. Capacitor matching errors can result from undercutting during the etching phase or from gradients arising from non-uniform oxide growth (Suárez *et al* 1975). Fortunately these problems can be minimised by choosing proper geometries and improving the oxide growth technique.

The precision-ratioed MOS capacitors allow the realisation of circuits such as precision analog/digital converters and audiofrequency filters (Hodges and Gray 1978) using all MOS analog sampled-data techniques.

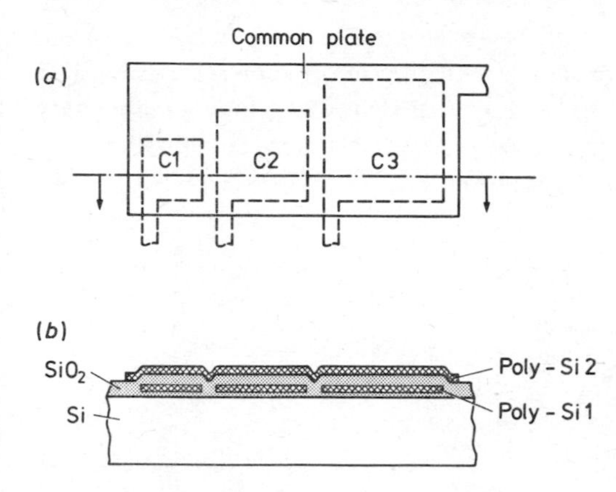

Figure 13. Precision-ratioed capacitor array in double polysilicon technology: (*a*) layout; and (*b*) cross section.

Table 5. Performance of a MOS operational amplifier.

Open loop gain	61 dB
Power dissipation (±5 V)	3 mW
Unity gain	13 MHz
Slew rate (+/−)	$20\ \mathrm{V}\ \mu\mathrm{s}^{-1}$
0·1% settling time	$<1\ \mu\mathrm{s}$
Equivalent input noise (RMS) (10 Hz–10 kHz)	$11\ \mu\mathrm{V}$

In conjunction with precision-ratioed MOS capacitor arrays nearly zero-offset analog switches are required. Fortunately the MOS transistor performs as an almost ideal analog switch when associated with on-chip capacitors. In a single-channel MOS circuit, the dynamic range of the MOS transistor is limited by one threshold drop below the supply voltage. If this becomes a serious limitation either bootstrapping techniques to increase the gate voltage or a dual-channel technology (CMOS) can be used and the analog switch realised by an n- and p-channel transistor connected in parallel.

So far only basic circuit elements and no circuit building blocks have been described. Because of its utmost importance to analog circuits the state of the art of the most important analog building block, the operational amplifier, will be reviewed.

Compared to bipolar technology MOS yields matched pairs of transistors with higher offset voltage and much lower transconductance. The equivalent input noise is also much higher, particularly the $1/f$ component caused by fast surface states. Despite these inherent drawbacks the innovations in operational amplifier designs have been quite remarkable. Table 5 shows an example of a recently described amplifier (Tsividis 1979) implemented in a single-channel technology. In the future this performance may even be further improved if a dual-channel technology is used. More important, however, is the technological compatibility of the operational amplifier which has been achieved with other MOS circuit elements, as this opens the way to many complex analog and analog/digital systems.

Considering the wide variety of MOS circuit elements available it would be desirable for the semiconductor producer to implement all circuit elements in one technology only. This would give the circuit designer the maximum design flexibility for all kinds of systems. Unfortunately it would also result in a most complex and therefore uneconomical process, and compromises must be worked out. This is illustrated in figures 14 and 15, where modern n-channel processes are shown with their most representative circuit elements and product lines respectively. It is obvious that figure 14 is only a momentary picture which changes as processes with different circuit element combinations are required or new and improved circuit elements are invented.

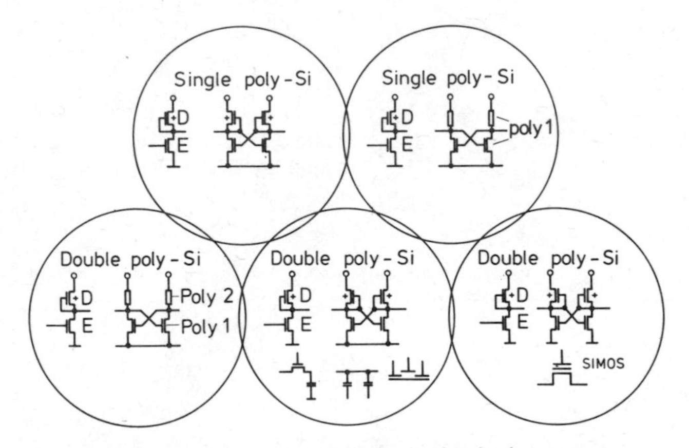

Figure 14. N-channel processes with basic circuit elements.

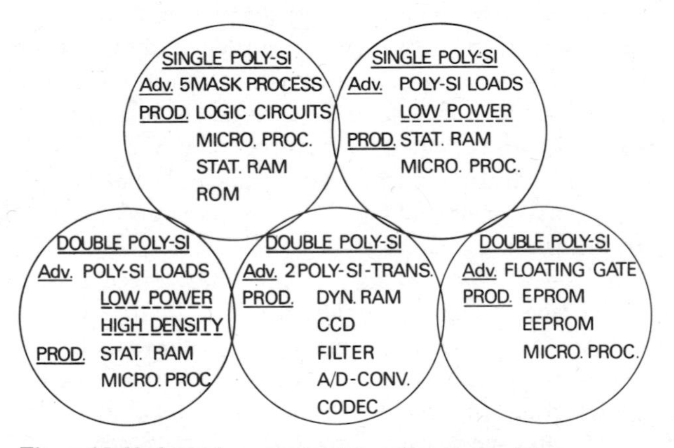

Figure 15. N-channel processes with various product lines.

4. Future developments

As has been stated previously device development will be governed in the forseeable future by the *scaling rate* of today's conventional n-channel technology. The major technological problem which has to be overcome is the generation and control of small feature sizes.

Indirectly generated MOS devices may gain in importance for very small dimensions as some restraints encountered by scaling a conventional MOS transistor can be overcome more easily.

In the past circuit element development was concentrated entirely on digital circuits. Today additional effort is made on the development of analog elements. In the future this development may be enhanced substantially by implementing, for example, audio-frequency filters using all MOS analog sampled-data techniques and precision analog/digital converters on one chip. The double polysilicon technology offers the best solution for this requirement and also offers the attractive opportunity of including novel electrically programmable analog and digital circuit elements on the same chip. So far the outlook seems quite clear compared to the uncertain future of the MOS industry, which stems from the dilemma of what to do with the gigantic number of components available on a VLSI chip. During the 1970s the MOS industry tried to cope with the complexity problem by designing memories and microprocessors (Moore 1979). This solution also applies partly for the 1980s, but so far only memory products fully utilise the available IC complexity (figure 16).

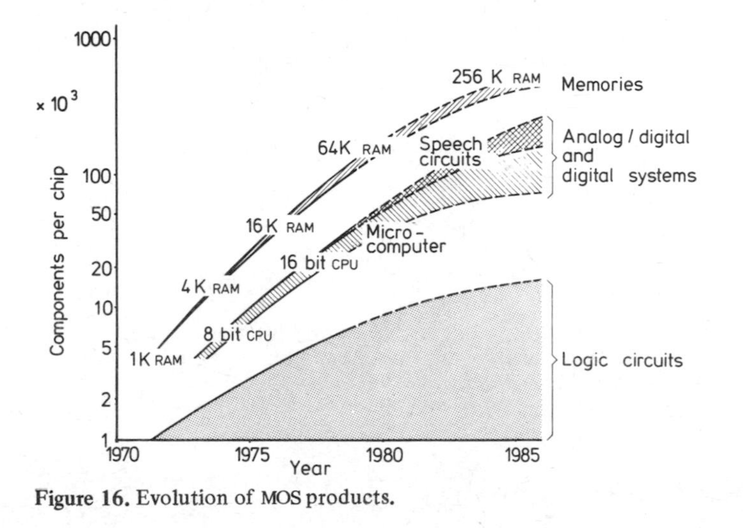

Figure 16. Evolution of MOS products.

It can be hoped that in the future the product definition problem may be overcome by the ability to implement digital and analog circuit elements on one chip. New circuits such as speech analysis, recognition, storage and per-channel PCM voice encoder–decoder, which serve as an entry into the area of telecommunication circuits, seem possible.

5. Conclusions

While the effort continues in the direction of scaling existing technologies for future VLSI circuits, the following conclusions can be drawn:

(i) For the near future up to about 1985 the scaling rate will be determined primarily by the progress in photolithography to reduce dimensions.

(ii) The introduction of new products will be governed mainly by the circuit designer's ingenuity to use the available circuit elements for new and useful high-volume VLSI circuits and his ability to cope with the gigantic number of components on a chip.

References

Dennard R H *et al* 1974 *IEEE J. Solid St. Circuits* **SC9** 256–68
Essel D V *et al* 1979 *ISSCC* vol XXIII (New York: IEEE) pp 148–9
Garbrecht K and Stein K-U 1976 *Siemens Forsch. Entwickl. Ber.* Bd 5 Nr 6
Hagiwara T *et al* 1979 *ISSCC* vol XXII (New York: IEEE) pp 50–1
Heil O 1935 *British Patent No.* 6815/35
Hodges D A and Gray P R 1978 *IEEE J. Solid St. Circuits* **SC13** 285–94
Hoffmann K and Losehand R 1978 *IEEE J. Solid St. Circuits* **SC13** 5 617–22
Lilienfeld J E 1930 *US Patent No.* 1 745 175
Lyman J 1979 *Electronics* pp105–16
McCreary J L and Gray P R 1975 *IEEE J. Solid St. Circuits* **SC10** 371–9
Mitterer R 1979 *Nachrtech. Z.* Nr 6 Bd 32 375–81
Moore G 1979 *IEEE Spectrum* April pp 30–7
Pashley R D *et al* 1979 *ISSCC* vol XXII (New York: IEEE) pp 106–7
Poon H C *et al* 1973 *Int. Electron. Devices Mtg Dig.* (New York: IEEE) pp156–9
Rodgers T J *et al* 1974 *IEEE J. Solid St. Circuits* **SC9** 5 239–50
Rodgers T J *et al* 1975 *IEEE J. Solid St. Circuits* **SC10** 5 322–31
Rössler B and Müller R G 1975 *Siemens Forsch. Entwickl. Ber.* Bd 4 Nr 6
Rideout V L 1978 *IEEE Workshop on Large Scale Integration, Stanford University, California, April 1978*
Rüchardt H 1978 *Elektrotechnik und Maschinenbau* 6/7
Siemens AG 1979 *Marketing Research*
Suárez R E *et al* 1975 *IEEE J. Solid St. Circuits* **SC10** 379–85
Tarui Y *et al* 1971 *J. Jap. Soc. Appl. Phys.* suppl. **40** 193–8
Tasch Al F Jr *et al* 1976 *IEEE J. Solid St. Circuits* **SC11** 575–85
Tihanyi J and Widmann D 1977 *Proc. Int. Electron. Devices Mtg, Washington* (New York: IEEE) pp399–401
Tsividis Y P and Fraser D L Jr 1979 *ISSCC* vol XXII (New York: IEEE) pp 188–9
Wallmark J T 1975 *Solid State Devices 1974* (Inst. Phys. Conf. Ser. No. 25) pp133–67
Wang P P 1977 *IEEE Trans. Electron. Devices* **ED24** 196–204
—— 1978 *IEEE Trans. Electron. Devices* **ED25** 779–86
Weiss H 1975 *Phys. Bl.* 31 Jahrgang Heft 4/5
Welker H 1945 unpublished
West J L 1979 *Microelectron. J.* 9 4

Medical applications of electronic devices †

H Thoma

Biotechnisches Laboratorium, c/o II Chirurgische Universitätsklinik Wien, Spitalgasse 23, A-1090 Wien, Austria

1. Bioengineering

More than 20 years ago, a new branch of science – bioengineering – was established as a result of the specialisation of medicine and technology; this new branch provided a connection between medical and technical science. 50 years ago, the physician developed new instruments with the assistance of a mechanic, while today, things have become more and more specialised, so that even in bioengineering new branches are having to be introduced. The relationships between medicine and biology and physics and technology are often misunderstood (figure 1*a*), for there appears to be a sharp separation between biomedical engineering, medical physics and biophysics. Some people believe that it depends very much upon the field from which an engineer or physicist is about to enter biology or medicine. In practice, however, this is not so, the situation being better depicted in figure 1(*b*). There is a long and mutual course during which the differences in the origins of the bioengineer or biophysicist vanish. There is no doubt that a difference exists between theoretically and practically oriented work, but on the other hand all branches experience the same sensation by working with living subjects. I want to point out another factor in the combined medical and technical efforts, namely, bioengineering can be viewed in industrial or clinical ways. As a result of my 15 years work at a University Clinic, I represent the latter (Thoma and Wolner 1972, Thoma 1975).

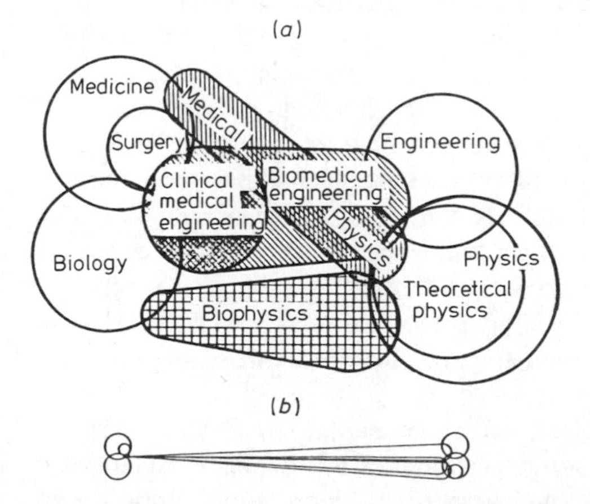

Figure 1. Interdisciplinary relationships between medicine and technology: (*a*) distorted; (*b*) natural.

†Supported by the Ludwig Boltzmann Institute for Cardiosurgical Research and the Austrian Research Funds.

0305-2346/80/0053-0097 $03.00

In what follows I will try to demonstrate not only the functions but also the professional specifications of the clinically oriented medical engineer. Before that, however, I would like to discuss the importance of electronics in medicine.

If someone feels ill, they probably would not consult a physician immediately; it is more likely that they would take some kind of medicine first. These medicines would not exist without the massive utilisation of electronic analysis. Today, pharmacology requires a vast amount of analysis and control instrumentation in preparation development, testing, and production. If someone has a serious disease – which is not to be hoped – and it is necessary to go to a hospital, for those who do not want the application of modern semiconductor technology there is only the chance of going to an old provincial hospital with a very good physician, who may eventually find a cure. In an average modern hospital, however, each department specialised to some extent usually begins with a *thorough examination.* During this examination several simple technical devices such as ECG, blood pressure gauge, x-ray devices, etc, are employed and more complex measurements might also be made, for instance nuclear diagnosis, ultrasonic measurements, cine-angiography, computer-aided tomography, and so forth. But much greater expenditure is needed for things that a patient will *not* see, for instance, the laboratories. A blood sample undergoes 30–50 different tests, and similar figures are achieved for other body liquids and samples, which are sometimes taken without the patients knowledge of how or why. A patient is then monitored, even before the therapy actually begins, and this monitoring of the vital functions is by electronic means. During therapy electronics play a less important role, except in certain special treatment methods which I will deal with later on.

2. Applications and criteria of electronic instrumentation in medicine

The applications of electronic instrumentation in medicine are certainly comparable with those of industry. Electronic aids are used for analysing, controlling, transmitting, displaying and computing biological, personal and statistical data. Requirements vary, depending on the actual application; for instance, diagnosis requires precision, therapy requires reliable function, laboratory work needs stability and automation, while patients' applications demand 'safety first'. A special application of bioengineering includes simple technical devices that aid personnel or the patient directly. Here is a simple example. Temperature measurements of a patient's body are made a couple of times a day. In Vienna several hundred thousand measurements are done per day. A primitive thermometer is used, which from the viewpoint of modern technology could have been obsolete for at least 50 years. Today any manager has his own ball-point pen with integrated calculator, so why does a nurse not have a similar temperature gauge with which she can determine the body temperature within two seconds? As the nurse has to visit each patient twice, she could also save 50 per cent of her working time on taking the temperature. Perhaps this simple example may stimulate inventors to contribute to a better understanding and solution of these problems.

Let us now examine the criteria of electronic instrumentation. Firstly, it is very interesting that today the whole spectrum of semiconductor technology is employed in medicine. To my knowledge, there is no other branch of science which utilises semiconductor technology so universally. The reason for this is quite simple: in general, the purpose of technology is to improve and aid the function and reception of human beings, that is, man is able to move faster and see, hear, speak and sense better with the aid of technology. If this basic principle applies to healthy human beings, the more it is valid for handicapped ones. Therefore it is understandable that each hardware component is used in various forms and combinations in the field of medicine.

The next criterion, if we look at it through an engineer's eyes, is disadvantageous: the quantities of electronic components used in medicine are very small compared to those used in industry and in society generally. Unfortunately, the consequence is that medicine sometimes uses technologies not of yesterday, but of the day before. It is almost a matter of course that people have a remote control with touch-sensitive buttons for their colour TV sets at home but in medicine, where such devices would be desirable, for instance for hygienic reasons, these techniques are just being introduced. Therefore, during the last few years, one of the programmes I have specified for our laboratory is to find medical applications for existing, highly integrated and consumer-oriented electronic parts and devices. An example of this is the application of highly integrated circuits (for instance the SAA 1050, an IC by Siemens) in a remote control for handicapped persons (Hoyer *et al* 1979).

Severely handicapped patients need permanent care and support, even for the simplest of tasks. Depending on the degree of disease, it is possible to ease the patients' lives by using simple technical aids. A remote control system allows control of several household utensils, for instance, the door opener, TV and radio, telephone, and so on. A pulse-code modulated RF transmitter is used, the electronic design providing trouble-free and convenient operation. The various RF receivers can be mounted directly at the point of execution (figure 2). Previous systems were rather expensive and bulky and could not of course be afforded by the handicapped.

A disadvantage of the small series required in medicine is that the semiconductor industry does not show a significant interest in the development of medically oriented ICs. A typical example is the heart pacemaker. Despite the fact that the principle has

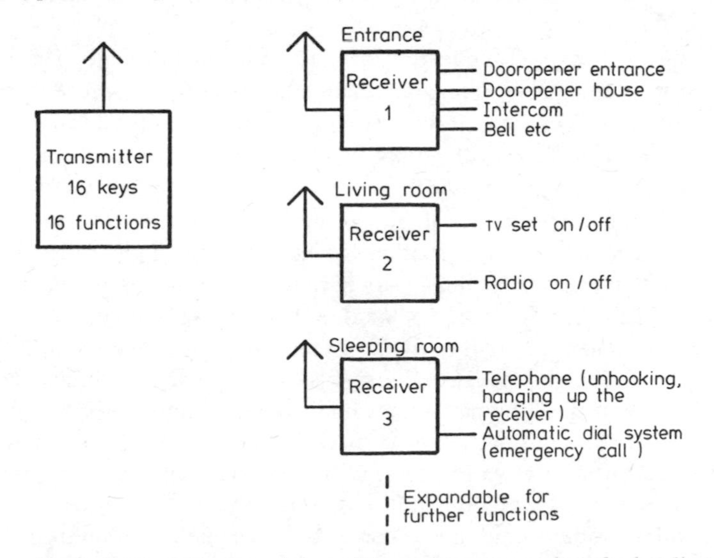

Figure 2. Block diagram of transmitter and remote receivers for handicapped persons.

been established for 20 years, only the last few years have seen some hybrid and higher integrated circuits becoming available to the physician. On the other hand, the turnover is not so small either. Each year our clinic implants pacemakers worth 3 million Austrian Schillings (about 400 000 DM). In my opinion there is an urgent need to reconsider the marketing concepts for medically oriented ICs. Firstly, the design costs can be reduced by having the work done by university laboratories. Secondly, the high cost of producing ICs in small quantities should not be carried by the patient, but rather by society. I do not see why, for example, governments provide funds only for research and development but not, to the best of my knowledge, for the production of medically oriented ICs. The benefits to society would in many cases demonstrate the profitability of integrated electronic circuits which would aid the rehabilitation of patients who otherwise have to rely on charity.

Another important criterion of medical electronics is the fact that a device is seldom built of electronic parts only. Even with simple devices, such as ECG amplifiers, the biotransducers (here electrodes) are more difficult to handle than complex electronic circuits. This leads to the statement that, especially in medical electronics design, the problems must be seen and solved as a whole. What good is the best electronic amplifier, combined with the best evaluation routines, if the electrodes are not positioned correctly or if faulty electrodes are used? Here is another important task for the medical engineer. It is not necessary for him to design the ECG amplifier, but he is responsible for the proper handling of the interface between instrument and patient. Another interesting observation is that, despite countless experiments, a trouble-free and reliable ECG pick-up method is still not available. The reasons for this are principally interface problems with skin electrodes and the recognition of pathological ECG waves.

Electronic devices in medical applications are usually combined with precision-made mechanical devices, such as equipment for laboratory diagnosis; organ substitution also depends on the high quality of mechanical devices. As I have said, the problems must be seen and dealt with as a whole, even if specific problems are handled by different technical disciplines.

The last criterion concerns the application of electronics in medicine. The medical instrument with its integrated electronic component usually serves its intended purpose well if, that is, the patient is brought into the laboratory. However, if the device is brought to the patient, great problems in overall system function are always experienced. A good example is intensive care monitoring, which can be described best by the term 'over-equipped technology' not only because of human deficiency, but also because a large percentage of machines connected to the patient do not function at all. Today, even the simplest ECG monitor may be faulty: a nurse may have to reset 99 alarms before the 100th shows a real heart malfunction. It can be understood then why various system alarms are frequently switched off. The same applies to the monitoring of blood pressure and other parameters. This lack of reliability leads to enormous extra cost and personnel expenses, not to mention the detriment of the patient. I cannot offer a solution for this problem. The structure of the relationships between the machine and the individual patient is so complicated that a great deal of research into this interface problem, both in theory and practice, would be necessary. I would include ergonomic studies from the outset, since in my opinion there is no other more unergonomic and uneconomic labour process than that observed in intensive care monitoring.

3. The status of electronics in medicine

It would be tedious and time consuming to list all the different applications of electronics in the many different medical disciplines; each of these branches of medicine has its own special electronic requirements. To obtain an idea of the amount of recent work undertaken in electronics in medicine, I analysed the papers presented at the 13th Annual Convention of the German Bioengineering Society in 1979. At this regional convention 24 topics were covered with about 10 papers per topic and involving 545 authors! For comparison, at the Annual Convention of the Austrian Bioengineering Society 6 topics were covered by 66 papers and 158 authors. These figures illustrate the vast amount of recent research being done in electronics in medicine. Regional differences do exist between various countries, but Eastern Europe exhibits similar figures. The numbers above show the importance of electronics in medicine from a quantitative point of view. Let us now examine it qualitatively.

4. Results of medical research aided by electronics

Research results in conjunction with modern semiconductor technology have a profound influence on modern medicine. In the following I shall try to give a few examples of the principal possibilities for medicine, made available through the use of electronic devices.

Many technical instruments today are used in non-invasive physiological measurement. The new medical standards were set by computer tomography, which is a display of the cross section of the body. The computer tomograph includes an x-ray unit, a computer and a monitor tube. The x-ray tube can be moved on a ring-shaped rail (figure 3), and the collimator, x-ray tube and x-ray detector are adjusted to each other precisely. The whole assembly can be moved in either an axial direction over the body or in circular motion around the body, which is positioned at the centre of the ring. Due to the axial and circular motion, the detector receives a three-dimensional density distribution, which is stored in the computer. By calculating the tube's position and the attenuation value caused by the body a three-dimensional matrix can be derived. A system of linear equations of the attenuation coefficients allows a quick computation of the density variation over a cross section, which is then displayed in a conventional manner. Figure 4 demonstrates the advantages of this method and shows cross sections through the human brain. Scan (*a*) shows the axial view, with a white line superimposed to indicate the coordinate at which the sagittal reconstruction was made. The position of this line is controlled by the tracker ball. Scan (*b*) is a vertical reconstruction showing the extent of the haemorrhage. Other organs, such as kidneys, liver, pancreas, etc, can be displayed and inspected with ease, and the equipment is invaluable for tumour diagnosis. Another great advantage over conventional methods is the accuracy of computer tomography, even without contrast agents.

The application of ultrasonics for non-invasive display of the inner body has been very successful in the last few years. Ultrasonic echo diagnosis utilises an ultrasonic pulse sent through the object to be investigated, in a similar manner to non-destructive testing of materials. At the boundary surfaces the pulse causes echoes which are received one after the other, due to their propagation time, by the same transceiver. The echo pulses are then displayed on an oscilloscope. This method is called 'A-scan' while a two-dimensional picture can be obtained with 'B-scan', where the transceiver can be moved laterally to

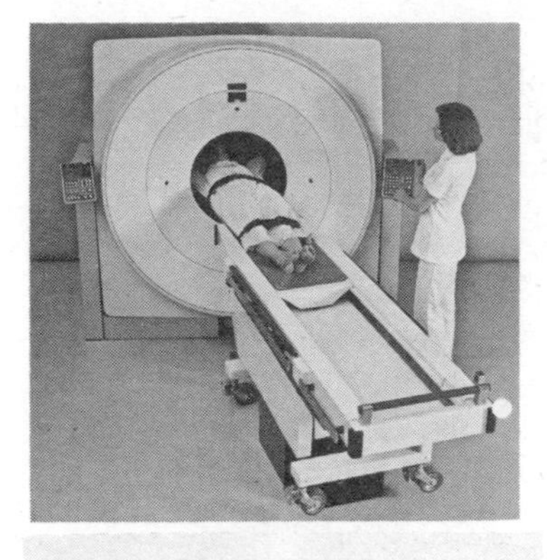

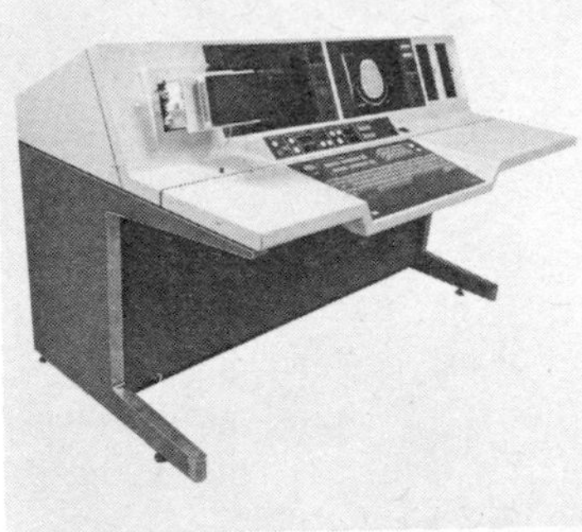

Figure 3. The complete EMI computer tomograph, x-ray and processing unit (display).

(a) (b)

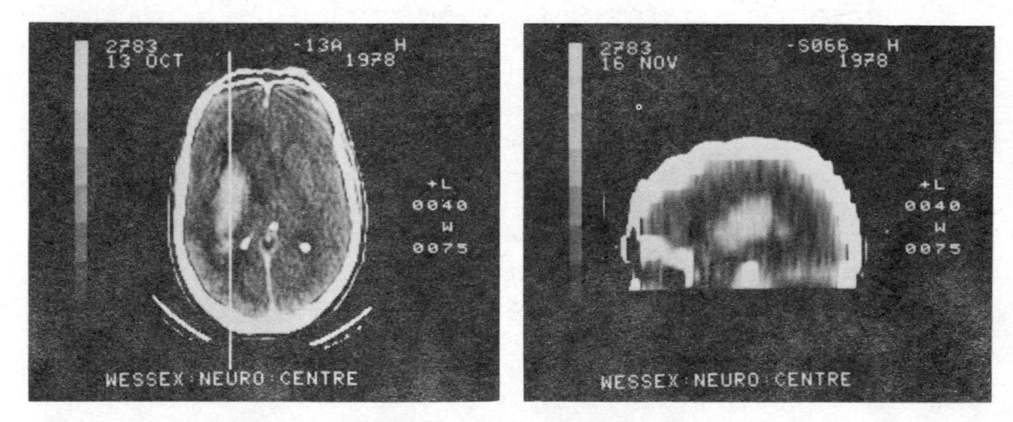

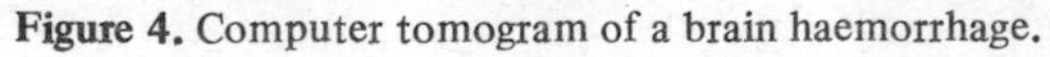
Figure 4. Computer tomogram of a brain haemorrhage.

the sound. Because air presents an extremely high resistance to ultrasonic sound, the object to be investigated must be acoustically well coupled to the transceiver. Scanners for greater surface areas use water containers or bags. The B-scan picture rate can be speeded up by using multiple sound transceivers, which leads to the so-called 'parallel scan' with oscillator array. With available funds, it is then easy to design a two-dimensional array which performs either A- or B-scans. The selection of echoes by their propagation time allows scans to be obtained diagonally to the body plane. The ultrasonic Doppler method is used to measure blood flow and velocity, and is possible because the blood particles reflect the sound (pure liquids would not give any ultrasonic echo). Latest designs include the quantitative measurement of the blood-flow velocity over the cross section of an artery. It may be of interest to note that instruments employing pulsed Doppler radar are still under development (figure 5) (Tsangaris *et al* 1979), but may be improved considerably by the application of cleverly designed surface-acoustic-wave filters for compression and decompression of the signal. Medicine distinguishes clearly between the display of the inner body and an organ's function. Therefore medical techniques provide separate methods for the representation of an organ's function.

Figure 5. Test assembly for ultrasonic flow meters (P Pfundner, Technical University of Vienna).

The function of an organ can be tested, for instance, by the use of radionuclides. Injection, oral administration or breathing allows the radioactive elements and chemical compounds to be transported into the body, where they accumulate in specific organs. If we inject, for example, radioactive particles of the right size to cause them to stick in the lung alveoli, the blood flow transports the particles only through the functioning areas. A gamma-detector, usually a photomultiplier, then scans these areas. The so-called 'scanner' does it serially by mechanical means, while the gamma-camera senses the whole investigation area at once. The latter contains a rather large number of photomultipliers, by which the spatial positions of the radioactive particles can be computed. Similar tests are available for investigation of the liver, kidneys, pancreas and other organs, and for thyroid investigation radio-iodine is utilised. This means that today almost non-traumatic organ function tests are at the patient's disposal, where function testing also includes dynamic activities.

So far these results of medical research have covered diagnosis only; therefore I would now like to consider the results in therapy and monitoring. It is known that physical medicine increasingly requires complex technical devices. For some years x-rays have been replaced by particle accelerators for malignant tumour treatment. Modern surgery would be almost impossible without the electrocauteriser, and today lasers are used routinely in cases of retina detachment, for cutting inner organs and in endoscopic surgery.

Some of the best results of medical research aided by technology are obtained in intensive care medicine. By definition, the intensive care patient requires the 'substitution of organ functions by technical means'. In intensive care units breathing is routinely controlled by machines and so are, to a lesser extent, blood circulation and kidney functions. Artificial feeding has special significance, since the balance of the metabolic process is usually disrupted severely by the disease. Such patients would then die of secondary effects rather than of the original disease. I still maintain my criticism about the function of technology in intensive care, but the medical success is manifested in various statistics. Due to the tremendous and expensive employment of personnel, material and equipment, approximately 65 per cent of intensive care patients can be cured. Despite justified criticism concerning the health service systems of socially oriented countries, the enormous advantage of this must be admitted. In Austria each intensive care patient receives first class treatment, but pays nothing, regardless of his social status. There is only one fear left: the day may come when a patient cannot be treated because of a lack of funds.

The achievements I have described up to now are at the disposal of a patient. Tomorrow's medicine depends on today's development and research. In § 5 I will discuss some results of research projects with which I am myself affiliated or of which I have intimate knowledge.

5. Examples of research in Austria

Quite a number of interesting research projects are associated with the Semiconductor Technology Laboratory at the Technical University of Vienna. The group under the engineers Erwin and Ingeborg Hochmair, in cooperation with the laryngologist Kurt Burian, puts its main efforts into electric stimulation of the inner ear in order to achieve hearing impressions for deaf persons having inner ear malfunctions. An eight-channel electrode, developed by the Hochmairs, is placed into the cochlea and causes selective stimulation of the remaining active sensory cells. The electrode is controlled by an eight-channel stimulation unit, designed in thin-film hybrid technology. This control unit is driven and powered by induction from an external control unit (figures 6 and 7).

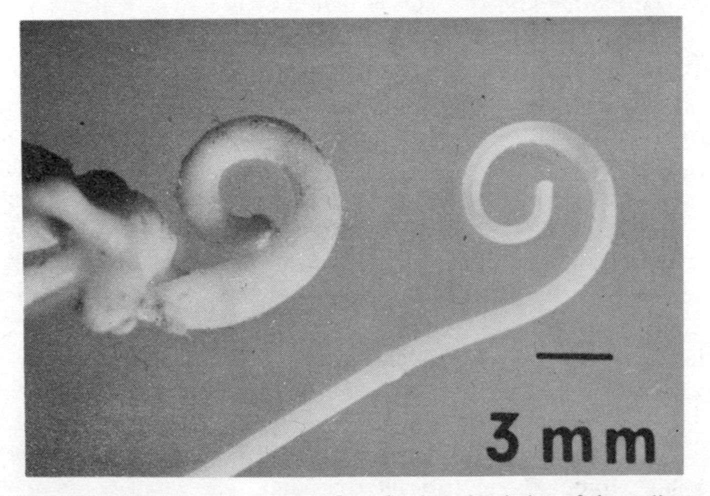

Figure 6. Eight-channel electrode for selective stimulation of the auditory nerve.

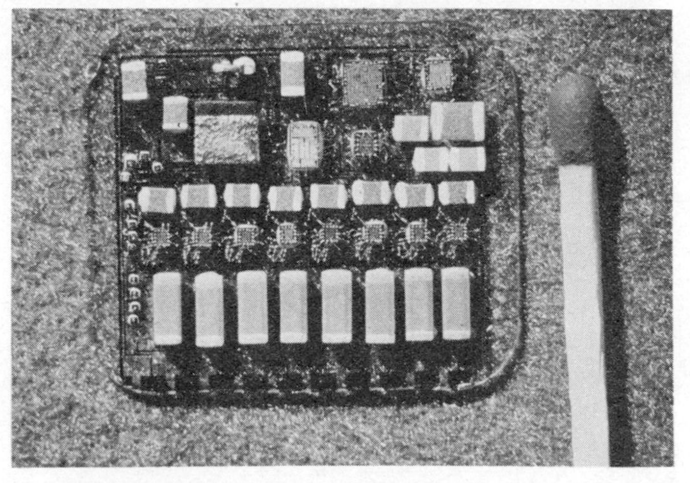

Figure 7. Implantable eight-channel stimulator, developed by E and I Hochmair (Technical University of Vienna).

New and essential perceptions in neurophysiology were revealed with a multi-channel electrode developed by Otto Prohaska (figure 8) and today this product of microelectronic thin-film technology is an irrevocable and necessary tool in numerous neurophysiological research projects (Prohaska 1975).

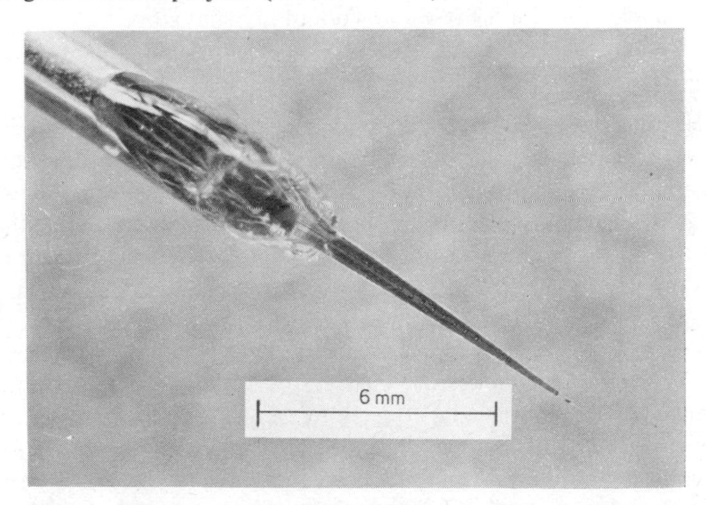

Figure 8. Multiple micro-electrode (O Prohaska, Technical University of Vienna).

From the various projects undertaken at the Technical University of Graz I have selected one that represents a new ultrasonic application, namely, the solid-angle scan method, which measures spherically the echoes of a ultrasonic signal. If the ultrasonic echoes are measured at different three-dimensional positions, reflecting boundary surfaces can be related to a characteristic spatial distribution of ultrasonic echoes. The quality of the transmitter can be easily tested with ideally smooth planes. If the transmitter

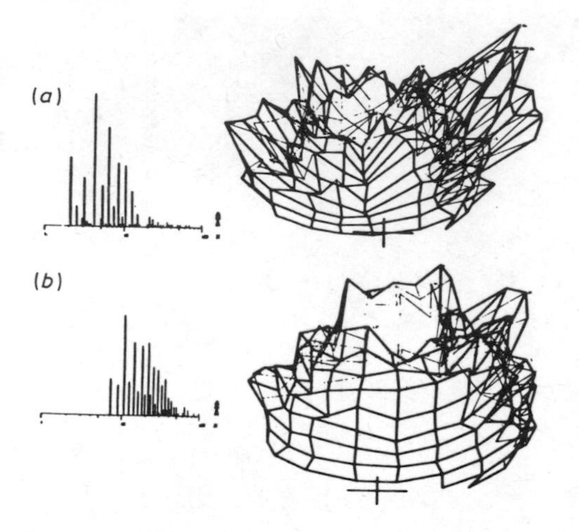

Figure 9. Solid angle scan of a normal (*a*) and cirrhotic (*b*) human liver *in vitro* using a three-dimensional display of the related frequency distribution of measured amplitude values (St Schuy, N Leitgeb, Technical University of Graz).

crystal characteristics and the frequency are known, objective surface structures can be determined. It seems possible that the method can be used for an increase in objectivity in preparation histology (figure 9).

The following examples concern the research work of my laboratory.

5.1. Perception contact analysis

The first example concerns psychometry. Let us consider the situation of actor and audience. The actor's speech is perceived and individually classified by the audience, which exhibits measurable reactions. The total amount of the audience's reaction then represents the mental activation the actor is able to cause. Now let us examine the block diagram showing the measurement of a psychical contact between the speaker and the audience (figure 10). The action of the performer is perceived and individually assimilated. The organ functions vary in a typical way which can be sensed by electronic transducers and then measured. Averaging over a representative group's reactions eliminates individual deviations. The result then shows that part of the action getting through to the audience (here we are talking about mental effects). Naturally, the expenditure on measuring instruments of such an installation is great, since a large number of parameters have to be monitored on each person, such as pulse rate, breathing rate, electric skin resistance, and so on. These parameters multiply with the number of persons to be tested. All data must be stored fully synchronously and the measuring instruments must not interfere with the situation itself. More than 10 years ago, such a transportable test set was developed, according to the criteria mentioned, and used for many years in various theatres, opera houses and concert halls in Vienna. The aim of the investigation was the *objective estimation of artistic performance.* The greatest problem was to find the key for the analysis of the sampled data, something which took 10 years to accomplish. Today we obtain 'biorhythmical function deviations' from the basic data. These deviations represent alterations of the vegetative nerve system. Data reduction is doen in two steps: the first

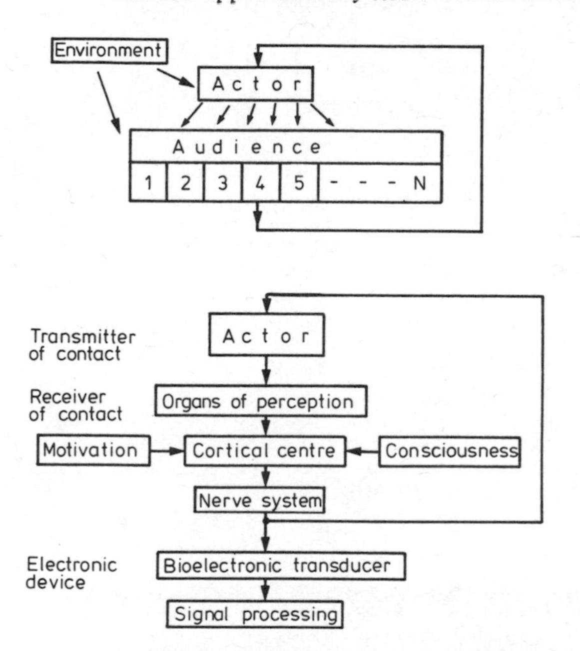

Figure 10. Block diagram of a sensibility test assembly in psychometry.

Figure 11. Data reduction of psychophysiological parameters (author's design).

in an analyser designed in our laboratory (figure 11), and the second in an overall evaluation done by a computer. A typical example (see figure 12) shows the total reactions of two test subjects measured during a performance of the Viennese Chamber Theatre. The abscissa shows the time in minutes and the ordinate shows the total reaction of one test subject referenced over 100 per cent. If we extrapolate this for a representative test group, a quantifiable intensity profile against time can be obtained for a whole stage play. These facts can be used in many ways in medicine, one example being the measurement of a patient's sensibility during surgery under acupuncture analgesia. During a research journey to the People's Republic of China in 1973, I had my first chance to collect objec-

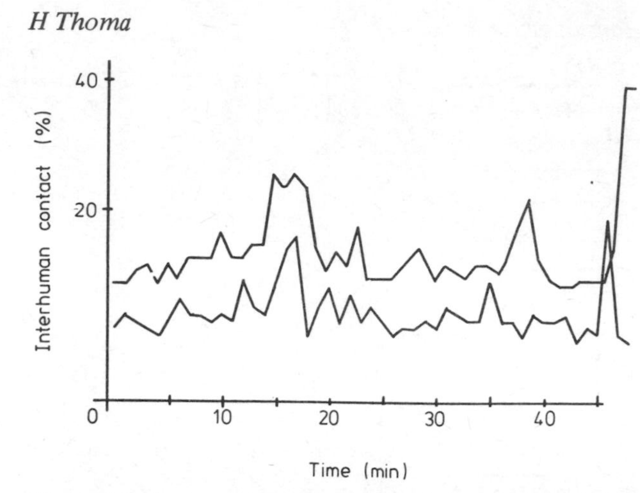

Figure 12. Graph of integral contact of two subjects during a performance of Jacques Duval's *Making Plans Is Everything* at the Chamber Theatre, Vienna.

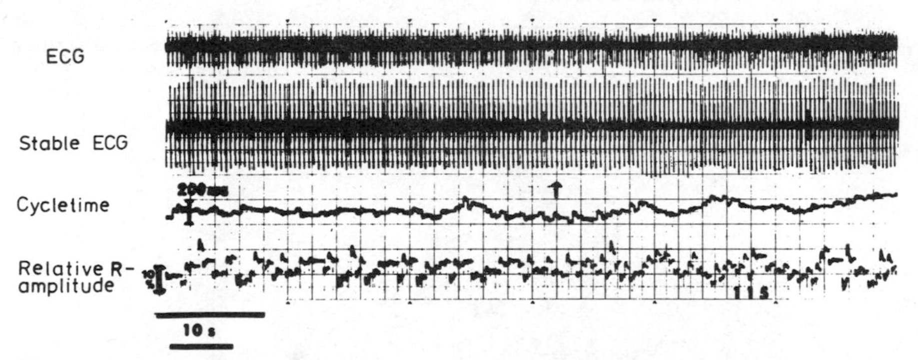

Figure 13. Oophorocystectomy under acupuncture analgesia. From top to bottom: ECG, stabilised ECG, period and R-amplitude deviation. The example shows the moment of skin cut; no reaction from the patient.

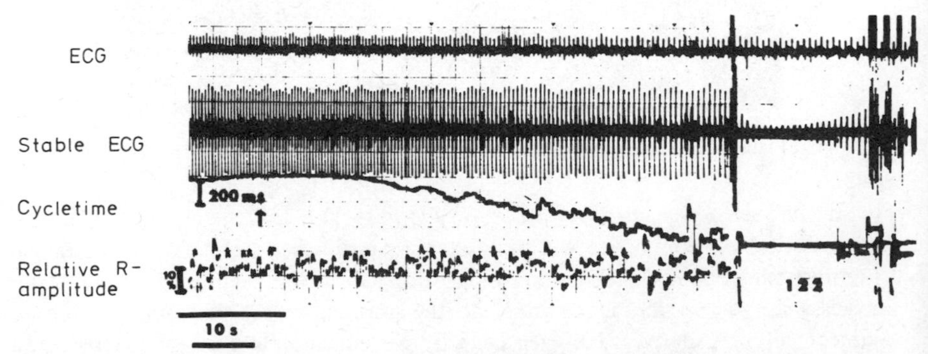

Figure 14. Lobectomy under acupuncture analgesia. Fear and/or pain experienced during the skin cut (see arrow). Parameters similar to figure 13, biorhythms appearing 2 minutes after skin cut. During skin cut, pulse rate is 120 per minute, decreasing to 60 per minute after 2 minutes. Interferences at the end of the figure are caused by electrocautery.

tive data with self-designed miniature transmitters (Thoma 1974). Two samples of this data collection are shown in figures 13 and 14. Both refer to the skin cut, a spontaneous and definitely painful event. In the first sample not even the biorhythmical changes of the heart action are lost, while the second shows a massive fear reaction, even before the skin cut, which then ceases two minutes after the event.

5.2. Functional electrostimulation

The next example describes a research project for functional electrostimulation after lesion of a cord (Thoma *et al* 1978). The increase in the number of these injuries is caused by the ever-growing number of sports and traffic accidents. The German Federal Republic alone accounts for more than 15 000 patients at the present time. Body areas below the lesion can no longer be controlled by the brain, though it is important to know that the patient's muscles, muscle nerves, and reflexes are still functioning. The brain's control over locomotion is replaced by electrical stimulation of the nerves. Figure 15 shows five electrodes implanted next to the nervus femoralis of a 25-year-old patient. During the experiment, which lasted four months, this patient was able to stretch her leg at the knee joint by means of a joystick. The aim of this project is clear: if we can functionally control the apparatus of locomotion for the lower extremities of paraplegic patients, it is certainly possible for them to stand and walk erect by means of (admittedly rather complex) electronic devices. This project will continue for some six to eight years before we achieve clinical applications on a broad basis. At present we are able to control the extensor of the knee and the flexor of the ankle. The technical department has designed an implantable eight-channel stimulation unit which receives control signals and power by means of an induction loop. Fatigue of the stimulated nerve, one of the main problems of electrostimulation, has been avoided by a new method (Thoma 1976), the so-called

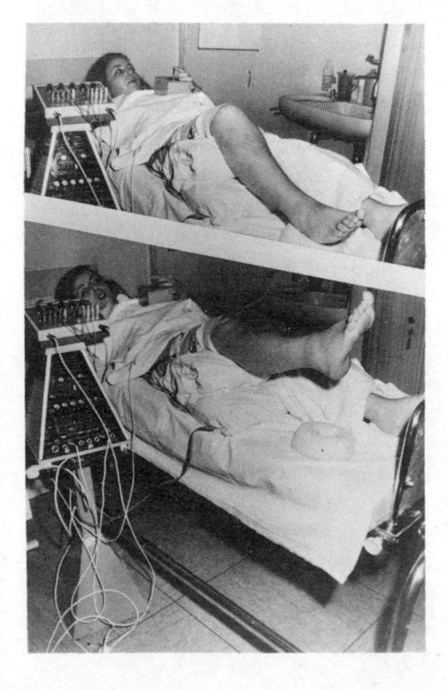

Figure 15. Functional stimulation of the skeleton muscle after lesion of the cord, extension of knee joint.

roundabout electrode. By using multiple electrodes positioned in a circle around the nerve and discharged in a random sequence, all nerve fibres of the whole nerve diameter can be equally strained.

As well as this experimentally oriented research work we routinely apply functional electrostimulation for central or peripheral breathing paralysis in the clinic (figure 16).

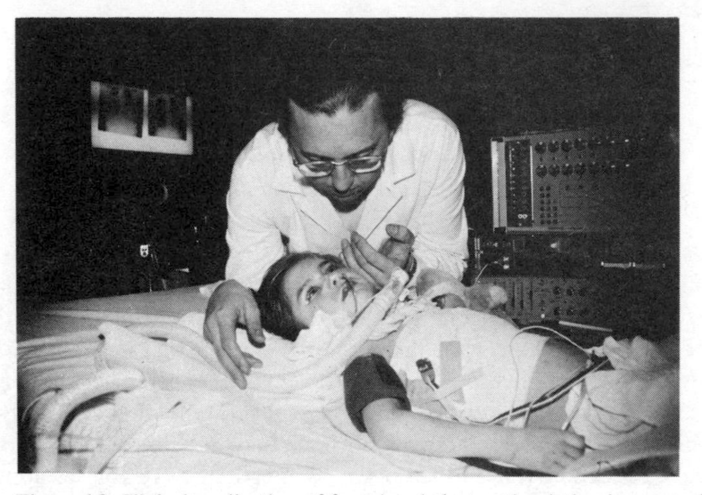

Figure 16. Clinical application of functional electrostimulation in a case of central respiratory paralysis.

This offers considerable advantages over the conventional method, with respirator aid 24 hours a day. With the conventional method patients need a tracheotomy and therefore cannot speak during respiration. For a group of patients with a lesion of the cord in the neck region we developed the so-called 'diaphragm pacemaker' (figure 17). The device consists of a implantable part, fed with control pulses and power through an induction loop, and an external unit controlled by a microprocessor. At present we are cooperating with the Semiconductor Technology Laboratory at the Technical University of Vienna to convert the discrete circuit into an integrated one. The miniaturisation of the external control unit will be the next step since ambulant support for paraplegic patients should be as lightweight and comfortable as possible.

5.3. Automatic control of artificial blood pumps

The last example of our own research work shows some results with automatic control of artificial hearts. There is no doubt that the substitution of human hearts with artificial blood pumps is one of the most exciting challenges in the history of medicine and technology. We have already been working for 10 years on the artificial heart project (figure 18). My main effort in this area is to design automatic control systems, and in the following a simple control system will be explained (Thoma *et al* 1976).

Today blood pumps are usually driven by pneumatic or hydraulic power, using the principle of volume displacement between the external pressure supply and the pump chamber. The volume displacement is transferred to the blood-filled part of the pump by means of a membrane; the blood flow is directed by special valves (figure 19).

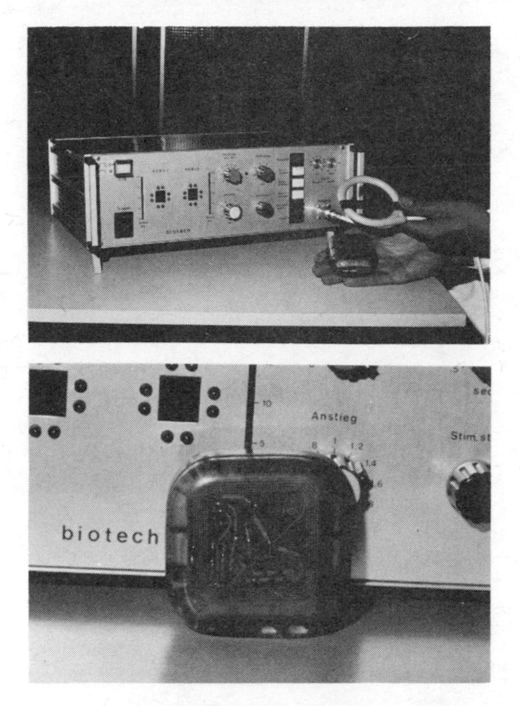

Figure 17. Implantable diaphragm pacemaker and external control unit.

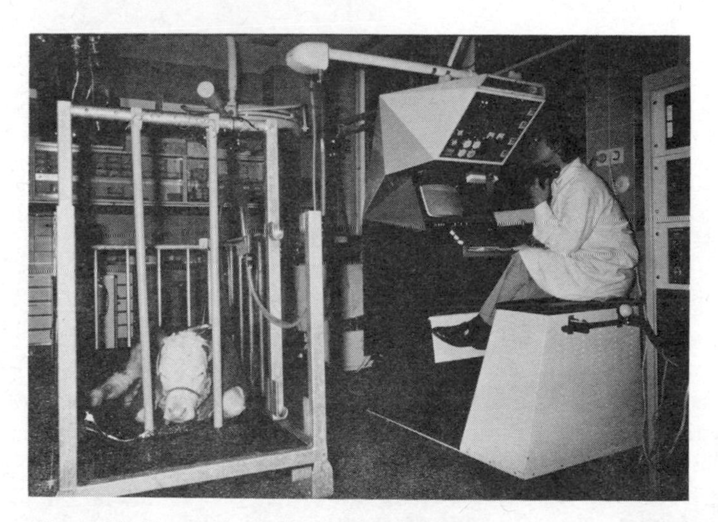

Figure 18. Total substitution of a calf's heart during an animal experiment. After basic studies, the automatic control is developed further using animal experiments.

Blood ejection is the result of a positive gas flow towards the pumping chamber, while blood filling is due to a negative flow. When switching occurs, the amplitude of the gas flow is a maximum. During the membrane movement the amplitude decreases, and at the moment of total ejection or total filling of the pump the flow amplitude ceases. By using a thermistor probe the output of the flowmeter is always positive, because the thermistor is cooled equally by positive and negative gas streams. We define total filling and total

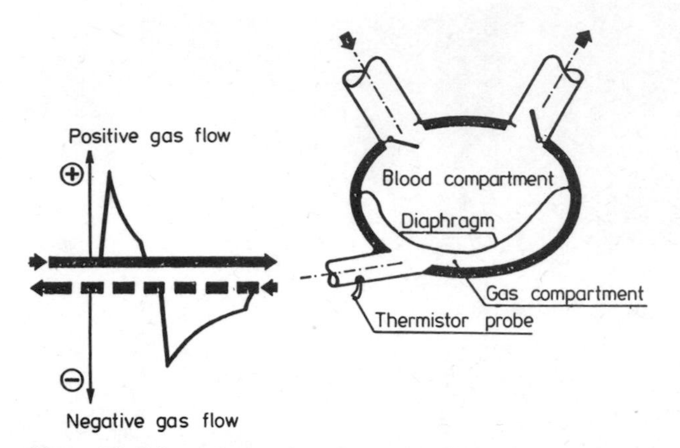

Figure 19. Schematic drawing of a pneumatic circulation pump drive.

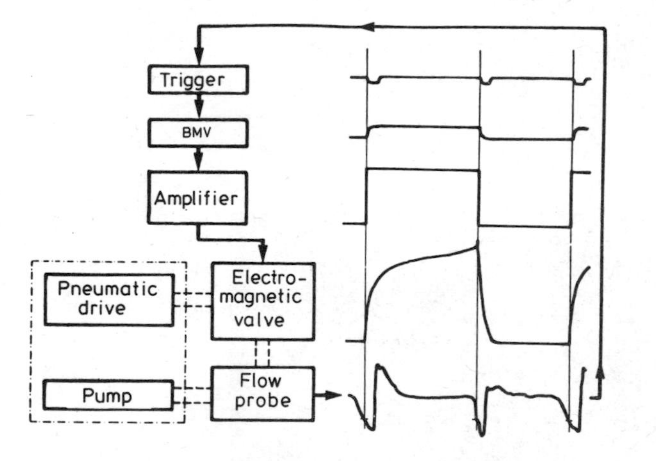

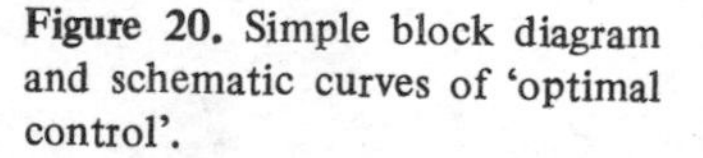

Figure 20. Simple block diagram and schematic curves of 'optimal control'.

ejection as 'optimal control' of the drive unit for articifial hearts (figure 20). A closed loop is utilised, beginning with the first return of the gas flow to zero, then continuing through the operation of a trigger and the reversing of the electromagnetic valve with a bistable logic switch. A typical registration of this 'optimal control' is shown in figure 21. Gas flow and pressure of the right and left pump are shown together with pulmonary and central aortic pressure. Note the continuous return of the left gas flow in the middle of the figure.

Filling and ejection times are functions of the input and output resistance, together with pressure differences between atrium and pump, and aorta and pump. Under normal conditions the pressure difference during the diastole is small and therefore the filling time is longer than the ejection time.

Naturally, the amount of electronic equipment needed is much greater in practice than shown in figure 20. For instance, it is necessary to start and monitor the closed loop. Despite the additional extras, such as the back-up generator for breakdown and a simple learning circuit for repeated breakdown of the closed loop, the technological progress

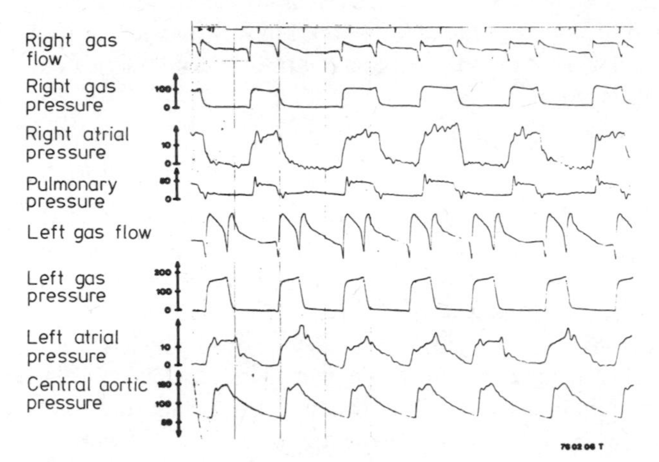

Figure 21. Typical 'optimal control' registration. Vertical axis, pressures in mm Hg, horizontal axis, time in seconds.

with semiconductor components allows the whole system to be put in a printed circuit board of EUROPE-format (100 x 160 mm). It is possible, but still not economic, to produce a one-chip version of this exclusively digital circuit.

6. Future aspects

(i) Therapeutic methods will be preferred to diagnostic methods, since patients do not want to be examined, but cured.
(ii) A computer substitution for the diagnosing physician, though rather remote, will be possible.
(ii) There will be a strong drive towards the rationalisation of products to overcome the economic problems of producing in small quantities.
(iv) The standardisation of diagnostic methods will be a major target.
(v) Unconventional therapeutic methods, for instance by following psychosomatic principles, will be developed.
(vi) A strong drive towards the adaptation of technical instruments and processes to the patient's needs will take place.
(vii) Improvement of established methods of analysis as regards function, long-term stability, automation, precision, handling and, last but not least, safety, is a certainty.
(viii) In semiconductor technology, the design and application of improved semiconductor transducers, such as flow, pressure and temperature gauges, improvement of existing and development of new biochemical sensors, e.g. for gases and glucose, are to be expected.
(ix) In therapy, organ substitution will surely play an important role. It can be clearly foreseen that the application of technical devices aiding the ear, eye, tactile senses, kidney, pancreas, lung and heart will only be restricted by financial resources. Special support devices for the cerebral and peripheral neurogenically damaged patient (for instance, in cases of epilepsy or paralysis) seem to be within reach of future semiconductor technology.

(x) Surgery in general will receive a new generation of manipulative devices (today surgeons still work with needle, thread and scissors). Fully automated appendix surgery does not seem to be possible yet, but computer-controlled manipulative devices are a natural consequence of technical progress.

7. Summary

Medicine uses all the results from modern technological research. It is significant though that electronics are almost never applied exclusively, but rather in combination with other technologies, mostly precision mechanics. Compared to consumer electronics, medicine experiences the disadvantages of small series.

The effects of modern semiconductor technology in medicine are enormous, and therefore we can say that without this technology modern medicine would not exist. We distinguish between laboratory and direct patient applications: the laboratory uses semiconductors as an aid to analysis; direct patient applications have inherent complex requirements, which will be the subject of future research. The main efforts of the present work in analysis deal with the improvement of bioelectric transducers. Spectacular success has been achieved in organ and function substitution. These projects in modern medical research definitely depend on semiconductor technology progress, since most systems must be implantable. The application of hybrid, large-scale integration and microprocessor technologies for implantation is complicated by the restricted power consumption allowance for such devices.

For a critical overview it is necessary to realise that a patient is still forced to 'consume' technology due to his disease, something which does not arise in other semiconductor applications. It is worthwhile discussing the whether the ethical requirements for technical applications in medicine should be stated or not. Practically speaking, the term 'patient-adjusted technology', proposed by the author, gains in importance because of the great redundancy of modern medical diagnosis. This raises unnecessary costs for the patient, though it does not entail any additional advantages (Thoma 1977).

The future will reveal two ways open to semiconductor technology application: the first is to improve established methods of analysis, and the other leads to the development and application of improved semiconductor transducers and new biochemical sensors. Both ways are necessary and both can be justified, since they lead to better perception by the physician.

References

Hoyer F, Schmallegger H, Stöhr H and Thoma H 1979 *Proc. Austrian Biomed. Engng Soc.* **4** 230

Prohaska O 1975 *Proc. Austrian Biomed. Engng Soc.* **1** 93

Thoma 1974 *Eur. Congr. of Anaesthesiology, Madrid. Excerpta Med.* **IV** 83

—— 1975 *Proc. Austrian Biomed. Engng Soc.* **1** 5

—— 1976 *Austrian Patent No.* 334 342

—— 1977 *Österr. Krankenh. Ztg* 185

Thoma H, Holle J, Moritz E and Stöhr H 1978 *6th Int. Symp. on External Control of Human Extremities, Dubrovnik, suppl.* **6** 70 (Belgrade: Yugoslav Committee for Electronics and Automation)

Thoma H and Wolner E 1972 *Ann. Mtg Austrian Surg. Soc.* **13** 207

Tsangaris S, Pfundner P, Jokobowicz D and Leiter E 1979 *Proc. Austrian Biomed. Engng Soc.* **4** 85

Laser annealing of semiconductors

Emanuele Rimini and Salvatore Ugo Campisano
Istituto di Struttura della Materia dell'Universita, Corso Italia, 57 – I95129 Catania, Italy

Abstract. Some aspects of the annealing of implanted semiconductors by means of high-power lasers are reviewed. Both Q-switch pulses and CW-scanning systems are considered with emphasis on the different mechanisms involved during these two processes. The redistribution of the implanted dopants is considered also as a test of the occurrence of a liquid phase. For Si–thin metal film structures, formation of silicides has been observed after pulsed laser annealing.

1. Introduction

Ion implantation, as employed by modern semiconductor technology, consists of the introduction of energetic, charged atomic particles into a substrate to change the electrical properties of the target. It offers some advantages such as the precise control of the dose over the implanted area and a good uniformity with respect to the usual diffusion processes. In addition, depth and concentration profiles can be varied independently by changing the beam energy and dose (see e.g. Mayer *et al* 1970).

As a non-equilibrium process it may be used to introduce dopants into the solid at concentrations in excess of the equilibrium solubility. The ultimate dopant concentration that can be implanted is limited only by the sputtering yield and by the energy of implantation.

A major problem with ion implantation is the creation of damage during the slowing down of the projectiles in the target. The near-surface layer of the implanted region is heavily damaged and for suitable combinations of projectile mass and energy, target mass and temperature it can be driven to become amorphous. It is then necessary to remove the lattice damage and to activate electrically the implanted dopant. The crucial step in the implantation process is concerned with the annealing.

So far almost exclusively thermal annealing has been used to remove the damage. For a partially damaged layer the high concentration of available vacancies makes defect annihilation possible and at the same time dopant atoms can occupy lattice sites and be activated electrically. For amorphous layers the system reorders in a solid-phase epitaxial growth (Csepregi *et al* 1976, Gyulai and Revesz 1979).

In many cases thermal annealing is not completely satisfactory. Extended defects such as dislocation loops remain in the light-ion implanted samples. Amorphous layers in (100) substrates reorder practically free of defects, while for substrates of (111) orientation, residual disorder in the form of twins is left. Dopants can also influence regrowth kinetics and damage removal. Precipitation of dopants in the implanted region and degradation of

0305-2346/80/0053-0115 $03.00

certain electrical properties occur after thermal annealing. In compound semiconductors, like GaAs, a protective cap must be used to avoid surface decomposition in the near-surface region with a loss of stoichiometry.

In these last few years new methods of annealing have been studied with the aim of overcoming the limitations associated with the thermal procedure. High-power laser irradiation, both in the Q-switched and in the continuous modes, together with electron beam irradiation are currently under investigation in many laboratories.

The interaction of a high-power laser beam with solid matter is not of course a new field of research. The idea of using laser irradiation to reorder implanted or amorphous layers was first published by Soviet scientists (Khaibullin *et al* 1975, Shtyrkov *et al* 1975a,b, Gerasimenko *et al* 1975). Interest in the problem increased suddenly in 1977 after the first USA–USSR seminar on ion implantation and research in this field was initiated in many laboratories. In the last two years many international conferences and workshops were dedicated to the subject. The number of papers increases continuously and it is thus impossible to give a really up to date review; therefore not all the work on the subject is quoted. Moreover in the present discussion we consider only some of the aspects relating to high-power laser irradiation. In the following, we will describe in some detail experimental data obtained mostly by Q-switched and continuous mode lasers in the damage removal of ion-implanted samples. The two modes differ in time scale. The single pulse of a Q-switched laser has a duration of the order of 10^{-8} s and an energy density of the order of a few joules per square centimetre. Continuous mode lasers are scanned over the sample with a velocity of a few centimetres per second, have spots of diameter 10^{-3} cm and power of the order of 10 W. (Lasers in the picosecond pulse-duration region have seldom been used.) The two main irradiation modes are associated with different physical mechanisms involved in annealing. With nanosecond pulses the entire amorphous layer is involved in the transition and the process shows a threshold character. A relevant dopant redistribution over a few hundred ångströms occurs after irradiation.

In continuous mode irradiations the dwell times are of the order of a few milliseconds; processes occurring in the solid phase at temperatures just below the melting point explain the experimental data.

As one aspect of this discussion we describe the experimental data obtained mainly by channelling, $^4He^+$ backscattering and transmission electron microscopy. A simple model in terms of heat absorption and dissipation by thermal conduction in the substrate is used to interpret the results.

2. Q-switched irradiations

2.1. *Experimental conditions*

In the Q-mode the lasers used are mainly ruby ($\lambda = 0{\cdot}694\,\mu$m) and Nd:YAG or glass ($\lambda = 1{\cdot}06\,\mu$m or $\lambda = 0{\cdot}53\,\mu$m with double frequency crystals). The absorption process in semiconductors depends strongly on the wavelengths and on the material structure. For instance in the 0·5–1·0 μm range the absorption coefficient of disordered or amorphous Si is an order of magnitude greater than that of crystalline material. The duration of the pulse ranges between 10 and 100 ns, and the shape is approximated by a Gaussian curve.

For most experiments in semiconductors the required energy density is of the order of a few joules per square centimetre.

The spot of the beam depends on the laser and its dimensions can range from a few square centimetres to a few square micrometres. In the latter case it is necessary to scan the sample. The basic experimental set-up is very simple for single shot irradiation. The laser beam goes through a beam splitter, reflecting a small fraction of the energy onto a photodetector to measure light intensity and waveform. In single shot irradiation the main problem arises from beam non-uniformity; hot spots are usually found in ruby laser beams for large irradiated areas. Small beam spots require good overlaps and longer durations.

2.2. *Amorphous to single crystal transition*

An amorphous to single crystal transition occurs above a threshold energy density value which depends on the layer thickness (Foti *et al* 1978a, b). Below this threshold an amorphous to polycrystalline transformation occurs. The threshold character of the transition to single crystal is illustrated in figure 1. A silicon sample of (100) orientation, self-ion implanted at liquid nitrogen temperature to produce an amorphous layer 4500 Å thick,

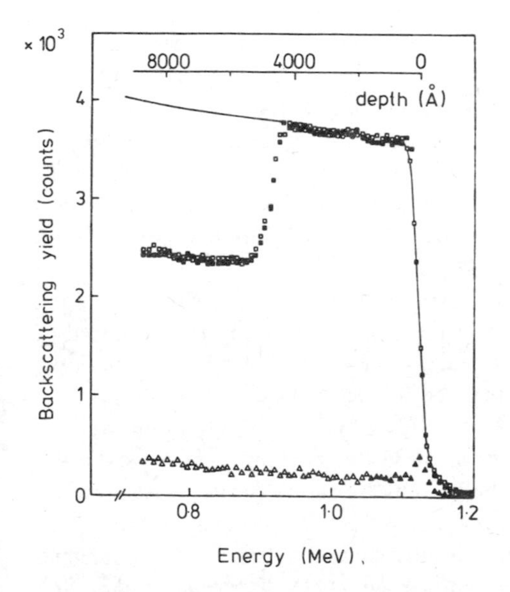

Figure 1. Random (full curve) and ⟨100⟩ aligned spectra for 2·0 MeV $^4He^+$ ions incident on an implanted sample (□), a sample after a 20 ns ruby laser irradiation of 1·8 J cm^{-2} (■) and 2·2 J cm^{-2} (△).

was irradiated with Q-switched ruby laser pulses of 50 ns duration and energy density ranging between 1 and 3 J cm^{-2}. The results of the channelling analysis using 2 MeV $^4He^+$ backscattering are reported in figure 1. The ⟨100⟩ aligned yield of the as-implanted sample reaches the random level for an energy width of 210 keV. The corresponding thickness of the disordered layer is about 4500 Å after the usual energy to depth conversion. The aligned yield after irradiation with a laser pulse of 1·8 J cm^{-2} coincides with that obtained from an as-implanted sample. After pulsing with an energy density of 2·2 J cm^{-2} the aligned yield decreases drastically and coincides with that of an unimplanted crystal. At this energy density the transition to single crystal occurs.

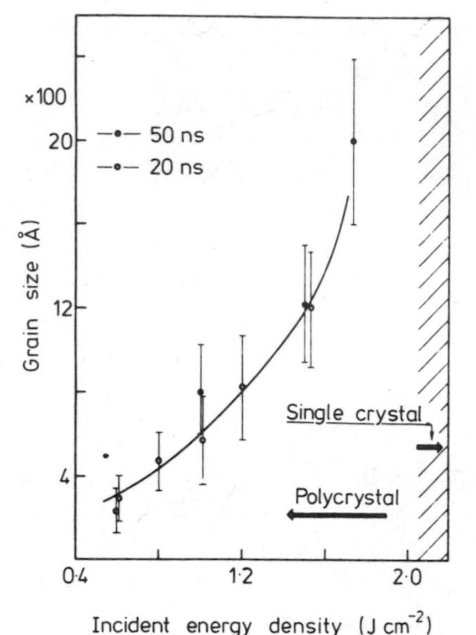

Figure 2. Average grain size measured by transmission electron microscopy plotted against the incident energy density for a 4000 Å thick amorphous layer on Si ⟨100⟩ illuminated with 20 and 50 ns ruby laser pulses.

Channelling techniques cannot distinguish between amorphous and randomly oriented poly-layers. To investigate the processes occurring below threshold it is then necessary to use another technique, e.g. transmission electron microscopy. Below threshold the amorphous layer becomes polycrystalline, as shown by TEM and diffraction. The grain size increases with energy density, as illustrated in figure 2, and approaching the single crystal transition the rate becomes very large. The measurements reported in the figure were taken with 20 and 50 ns duration pulses respectively and the results are the same for both durations (Tseng *et al* 1978).

A dependence on the substrate orientation was also found in the irradiated samples (Foti *et al* 1978c). As an illustration, in figure 3 the channelling analysis is reported of ⟨111⟩ and ⟨100⟩ oriented silicon substrates overlaid with a 4000 Å thick amorphous layer obtained by self-ion implantation at liquid nitrogen temperature. The aligned spectrum of the ⟨111⟩ sample irradiated at 2·5 J cm^{-2} (see figure 3*b*) has a minimum yield of about 12% near the surface and increases to a value of 35% at the implant to crystal interface. A high concentration of defects should be present as residual disorder. At the same energy density the aligned yield of the ⟨100⟩ crystallised layer is comparable with that of an unimplanted Si sample. The ⟨111⟩ sample irradiated at 3·5 J cm^{-2} gives an aligned spectrum which coincides with that of an unimplanted ⟨111⟩ Si crystal (figure 3*c*). A detailed analysis by TEM has shown that the ⟨111⟩ crystallised layer after irradiation at 2·5 J cm^{-2} contains a large density of intrinsic stacking faults while in the ⟨100⟩ sample some portions exhibit screw dislocations instead. At higher energy densities both crystallised layers are practically free of defects. In a comparison between the two orientations one may affirm that both require nearly the same energy density for the transition to single crystal of the amorphous layer. A difference arises in the residual disorder: stacking faults and dislocation lines for the ⟨111⟩ and ⟨100⟩ orientations respectively. In any case a suitable energy density will allow a complete restoration of the perfection of the crystallised layers.

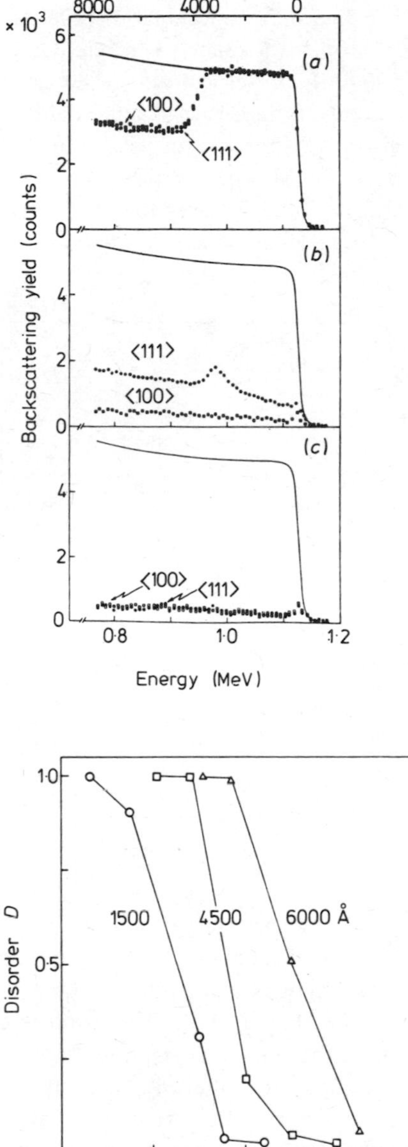

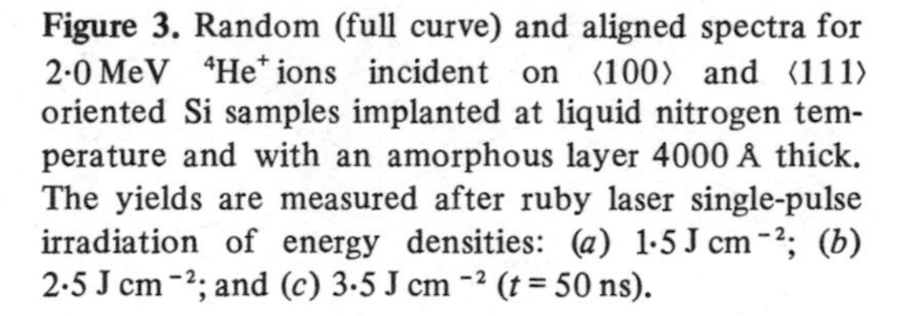

Figure 3. Random (full curve) and aligned spectra for 2·0 MeV $^4He^+$ ions incident on ⟨100⟩ and ⟨111⟩ oriented Si samples implanted at liquid nitrogen temperature and with an amorphous layer 4000 Å thick. The yields are measured after ruby laser single-pulse irradiation of energy densities: (*a*) 1·5 J cm^{-2}; (*b*) 2·5 J cm^{-2}; and (*c*) 3·5 J cm^{-2} ($t = 50$ ns).

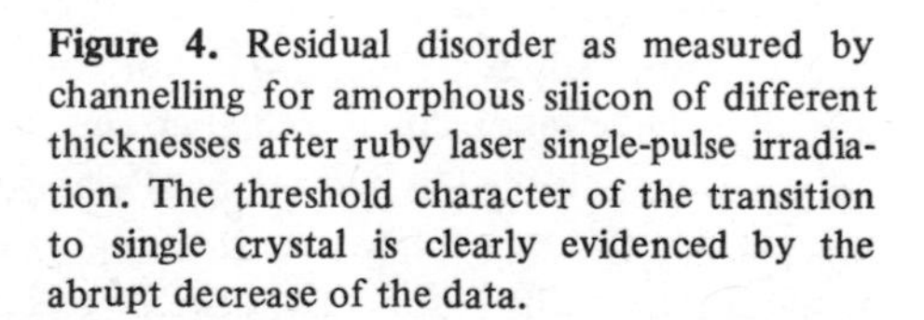

Figure 4. Residual disorder as measured by channelling for amorphous silicon of different thicknesses after ruby laser single-pulse irradiation. The threshold character of the transition to single crystal is clearly evidenced by the abrupt decrease of the data.

By changing the amorphous layer thickness a similar threshold behaviour was found as illustrated in figure 4. The residual disorder, as measured by channelling, decreases abruptly at an energy density which increased with the amorphous thickness. It amounts to about 1 J cm^{-2} for a 1500 Å thick silicon layer and reaches 3·0 J cm^{-2} for a 6000 Å layer.

The threshold value for the transition to single crystal depends also on the substrate material. Ge- (Foti *et al* 1978d) and GaAs- (Golovchenko and Venkatesan 1978,

Campisano *et al* 1978) implanted samples usually require a lower energy density for the transition to single crystal than is necessary for Si. In figure 5 the residual disorder, as measured just below the implanted layer in the aligned yield, is reported for Ge, GaAs and Si with amorphous layers 3000, 2300 and 1500 Å thick, respectively, after laser irradiation. Values of about 5–10% indicate a quite complete reordering of the damage layer. The sharp decrease in the damage level occurs for the three specimens at different energy density values. The transition to single crystal for Si requires a larger energy density than that for Ge although the amorphous thickness of Ge is twice that of Si. The ratio between the estimated energy density for Si and Ge melting is approximately two – in qualitative agreement with experiments.

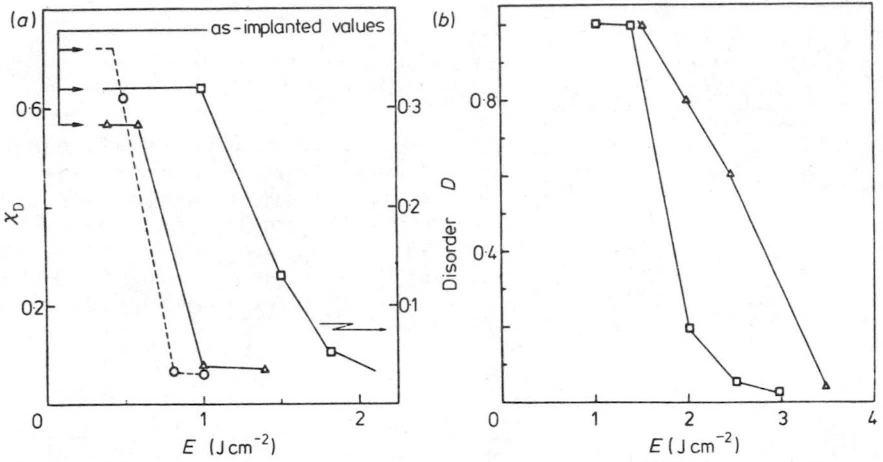

Figure 5. Residual disorder as measured by channelling effect for: (*a*) Ge (○), GaAs (△) and Si (□) ion-implanted amorphous layers of 3000, 2300 and 1500 Å thickness respectively; and (*b*) for 4500 Å amorphous (□) and polycrystalline (△) Si layers after ruby laser single-pulse irradiation (Q-switched, $\tau = 50$ ns).

The threshold value depends also on the structure of the absorbing medium as illustrated in figure 5(*b*), where the disorder measured by the channelling effect technique is plotted against the laser pulse energy density for amorphous and polycrystalline Si layers, both 4500 Å thick on ⟨100⟩ oriented substrates (Vitali *et al* 1978). The reordering of the poly-layer occurs at a higher energy density than that of the amorphous. At the ruby wavelength the absorption coefficient decreases by about one order of magnitude on going from the amorphous to the single crystal. Poly-layers are characterised by intermediate values which depend on the grain size and on disorder.

In summary, the experimental results for the annealing of disordered layers obtained by ion implantation in semiconductor materials indicate that the transition to single crystal is characterised by a threshold value which depends on the layer thickness, on the substrate material and on the structure of the absorbing region. It must be pointed out that crystallisation induced by Q-switched laser pulses is not confined to ion-implanted layers, as so far discussed. Several thousand ångstrom thick, vacuum-deposited silicon and germanium layers on Si substrate were crystallised by both ruby and Nd laser pulses, indicating the feasibility of homo- and heteroepitaxy by means of laser irradiation (Revesz *et al* 1978, Bean *et al* 1978).

2.3. Phase transition

As a first, simple approach the interaction of a laser beam with an ion-implanted semiconductor can be described in terms of energy transfer from the photon to the phonon field. The laser light is absorbed by electron excitation in the near-surface region of the solid. The excess of energy decays in a period of the order of 10^{-12} s by collision with thermal vibrations and free electrons. The absorbed energy is nearly instantaneously converted into local heat which diffuses by thermal conduction. The characteristic time for heat diffusion through a few thousand ångströms of material is of the order of 10^{-10} s. The laser pulse duration, of the order of 10^{-8} s, is in any case much larger than the diffusion time. The heat equation with a source term to include light absorption then provides a useful description. Assuming for simplicity a beam intensity uniform in cross section the problem becomes unidimensional and the heat equation assumes the following form:

$$c\rho \frac{\partial T}{\partial t} = I(z, t)\alpha + \frac{\partial}{\partial z}\left(k \frac{\partial T}{\partial z}\right) \tag{1}$$

where c is the heat capacity, ρ the mass density, α the absorption coefficient, k the thermal conductivity and $I = I_0(t)(1-R)\exp(-\alpha z)$ is the light power density inside the specimen and is a function of the time dependence, $I_0(t)$, of the incident light and of the reflectivity R.

A comparison with experiment requires a numerical solution of equation (1) to account for phase transitions and for changes of optical parameters (α, R) and thermal parameters (c, k) with temperature and structure. Typical results of the computation are shown in figures 6 and 7 and respectively. The time dependence of the temperature inside the sample was evaluated at several depths inside the sample ranging from 500 to 6500 Å and for a Gaussian pulse of 1·5 J cm^{-2}. The surface layer melts and remains liquid for about 130 ns. During the irradiation a maximum temperature of about 2000 K is reached for a duration of about 30 ns. The 2500 Å deep layer remains liquid for only 30 ns. For the layers located at 4500 and 6500 Å the material remains solid at all times,

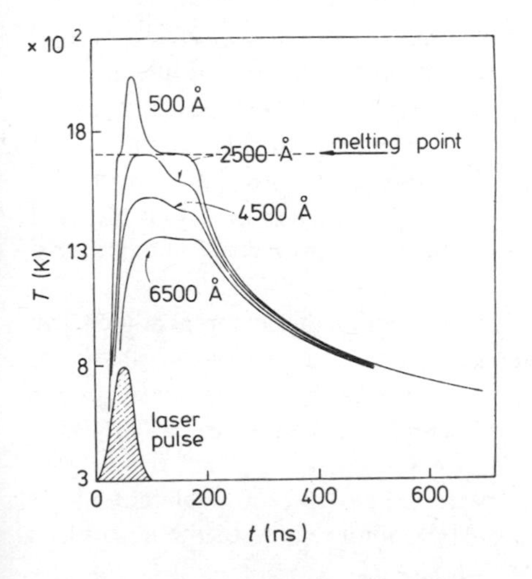

Figure 6. Calculated temperature–time profile at several depths inside the amorphous silicon layer. The energy density of the Gaussian ruby laser pulse is 1·5 J cm^{-2}.

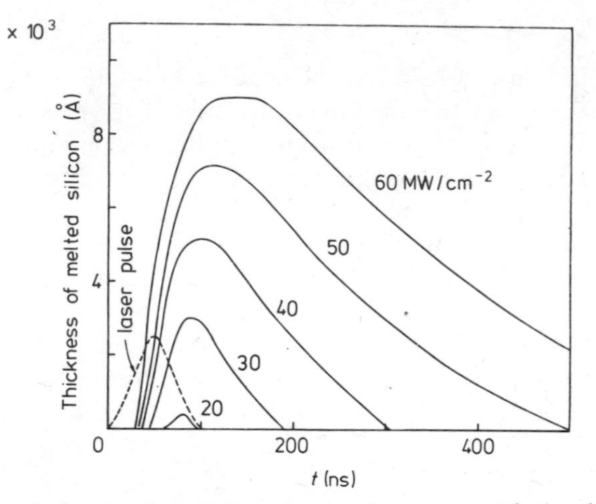

Figure 7. Time evolution of the solid–liquid interface for different power densities. The amorphous silicon melts above 20 MW cm^{-2} for 50 ns pulse duration. The interface moves toward the interior of the sample, reaches the maximum depth and then moves back to the sample surface.

and the maximum temperature decreases with depth. The cooling rate does not change appreciably with depth.

This depth analysis evidences the presence of two contiguous phases: a liquid layer extending from the surface to about 2500 Å, and a hot, underlying solid region. The solid–liquid interface can be followed as a function of the time and power density of the laser pulse. The results are shown in figure 7 where the thickness of the melted Si is plotted against time for several power densities ranging between 20 and 60 MW cm^{-2} and for a pulse duration of 50 ns. At 30 MW cm^{-2} the surface layer melts after about 45 ns and the thickness of the liquid layer reaches a maximum depth of 3000 Å at 90 ns. Thereafter the thickness of the melted layer decreases, the solid–liquid interface moves outward and the surface remains melted for about 150 ns. The maximum thickness of the melted region and the lifetime of the surface as a liquid phase increase with the power of the laser pulse. At 50 MW cm^{-2} a layer 7000 Å thick is melted and the surface remains liquid for up to 500 ns.

Results of the calculations are compared with experiment in figures 8(*a*) and (*b*) for Si and Ge samples respectively, on the basis that the transition to single crystal occurs when the molten layer is thick enough to wet the underlying single crystal. The curves in both figures plot the threshold energy density against amorphous thickness. Full squares indicate polycrystalline material, partially full squares represent residual disorder and open squares mark out the transition to a single crystal structure practically free of defects. The calculated curves divide the graphs into two parts each: below, the energy density does not allow the overall melting of the amorphous layer, above, the entire layer is melted. The overall agreement is quite good in view of the many assumptions involved in the calculation and of the experimental uncertainty related to power measurement and uniformity over the irradiated area.

Compound materials such as GaAs show a behaviour similar to elemental silicon and germanium. One advantage of Q-switched laser pulsing is related to the possibility of annealing GaAs without any capping. At least for Te implants the stoichiometry of the near-surface region as measured by channelling techniques was preserved (Campisano *et al* 1978).

In addition to ruby lasers, Nd:YAG or glass lasers are also used for semiconductor annealing. The main difference between ruby and neodymium relates to the absorption

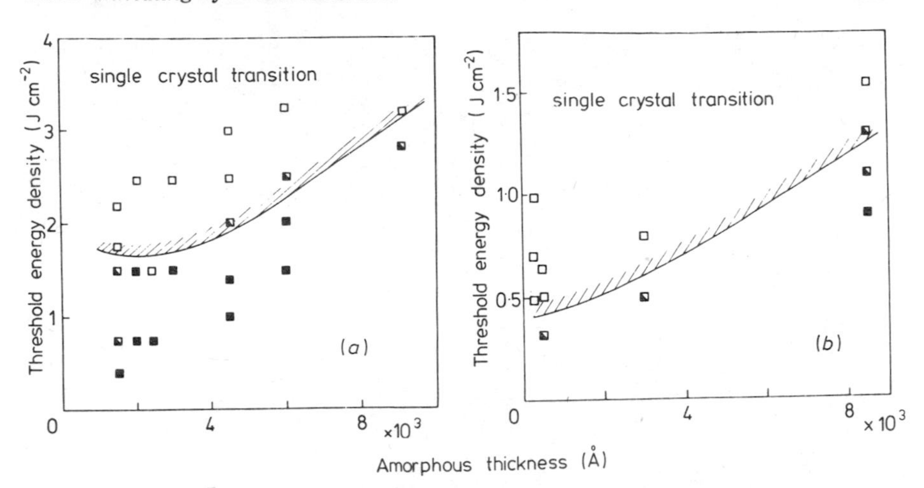

Figure 8. Structure of amorphous Si (*a*) and Ge (*b*) layers of different thicknesses after 50 ns ruby laser single-pulse irradiation. Full, open and partly full symbols indicate polycrystal, good single crystal and single crystal with high residual disorder respectively. Full curves plot calculated threshold energy density against amorphous thickness.

coefficient of Si at these wavelengths, and, hence, to the threshold energy density for melting the surface; $\alpha = 5 \times 10^4 \mathrm{cm}^{-1} (\lambda = 0{\cdot}69\,\mu\mathrm{m})$ and $\alpha = 2 \times 10^4 \mathrm{cm}^{-1} (\lambda = 1{\cdot}06\,\mu\mathrm{m})$, respectively, for amorphous silicon. As soon as the surface starts to melt the absorption coefficient becomes nearly equal for both wavelengths at approximately $4 \times 10^5 \mathrm{cm}^{-1}$, liquid Si being a metal. An increase by a factor of two in the energy density is required to obtain a single crystal transition with a Nd laser over that for a ruby laser.

During the solidification the cooling rate is about $10^9 \mathrm{K\,s}^{-1}$. The estimated solid–liquid interface velocity amounts to a few metres per second, with a temperature gradient of $10^6 \mathrm{K\,cm}^{-1}$. These values render this process very interesting because of the possibility of forming metastable phases.

2.4. *Impurity redistribution*

Irradiation with Q-switched laser not only removes the lattice damage in ion-implanted semiconductors, but also causes a relevant redistribution of the impurity atoms. In many cases the concentration of retained substitutional dopants exceeds the solid solubility by a few orders of magnitude. The matter transport caused by laser pulsing is related to the formation of a liquid layer. The final dopant profile depends on several factors: diffusion in the liquid phase, segregation coefficient, solid solubility limit, velocity of the solid–liquid interface during freezing, thickness and lifetime of the molten layer (Baeri *et al* 1978).

Figure 9 shows the effects of pulsed laser annealing on dopant concentration profiles for the cases of boron, phosphorus and arsenic implanted in silicon at doses of about $10^{16} \mathrm{cm}^{-2}$ (White *et al* 1978). Profiles for B and P were measured by secondary ion mass spectroscopy and the profiles for As were measured by backscattering. The redistribution of these dopants cannot be explained by diffusion in the solid state. The diffusion coefficient of As in Si just below the melting temperature is about $10^{-14} \mathrm{cm}^2 \mathrm{s}^{-1}$. The broadening of the profile is then negligible (<1 Å) for a time interval of 100 ns.

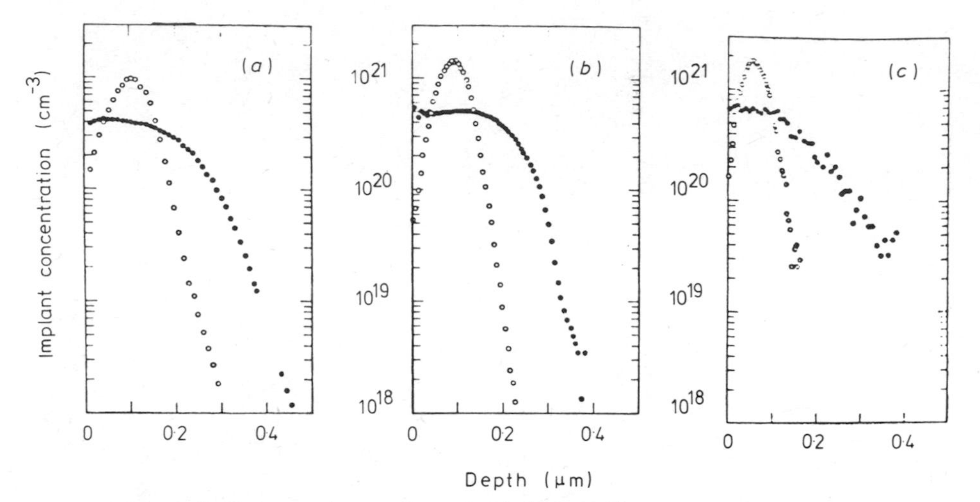

Figure 9. Concentration profiles of: (*a*) ^{11}B; (*b*) ^{31}P; and (*c*) ^{75}As implanted silicon. The profiles were measured before (○) and after (●) laser annealing.

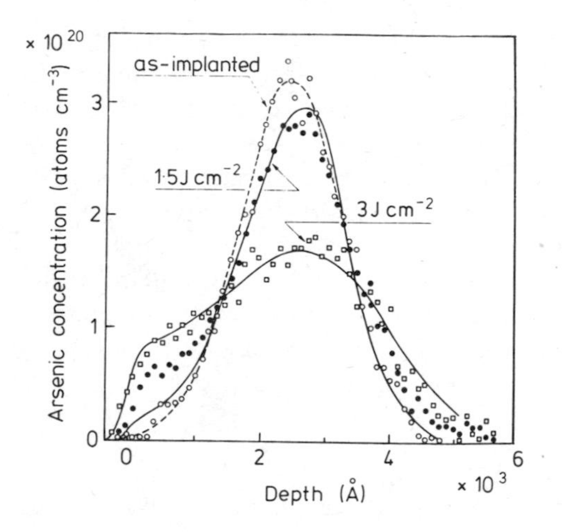

Figure 10. Arsenic profiles before and after laser annealing. Full curves are calculated assuming normal diffusion kinetics with a constant diffusion coefficient in the liquid state. 400 keV As, 5×10^{15} ions cm^{-2}.

Diffusion in the liquid phase, however, allows a good fit of the experimental data for the high value of the diffusion coefficient, which is of the order of 10^{-4} cm^2 s^{-1}. A comparison of arsenic profiles before and after laser annealing is shown in figure 10 with a theoretical profile calculated by the diffusional redistribution method. The full curve is the result of Gaussian diffusion analysis; using a time dependence of the liquid layer as illustrated in figure 7, the best fit is obtained for a diffusion coefficient $D = 10^{-4}$ cm^2 s^{-1}. Broadening of the impurity profile also depends on many factors, for instance the lifetime and extent of the liquid phase. To test the model, As ions at 40 keV were implanted in Si single crystals or in Si with an amorphous layer 4500 Å thick. The measured profiles are reported in figure 11(*a*) after irradiation with ruby laser pulses of differing energy densities for the two structures considered. The as-implanted profile is shown by a broken curve; curves labelled A and B represent the experimental profiles for the samples with 4500 and 500 Å thick amorphous layers respectively. The impurity re-

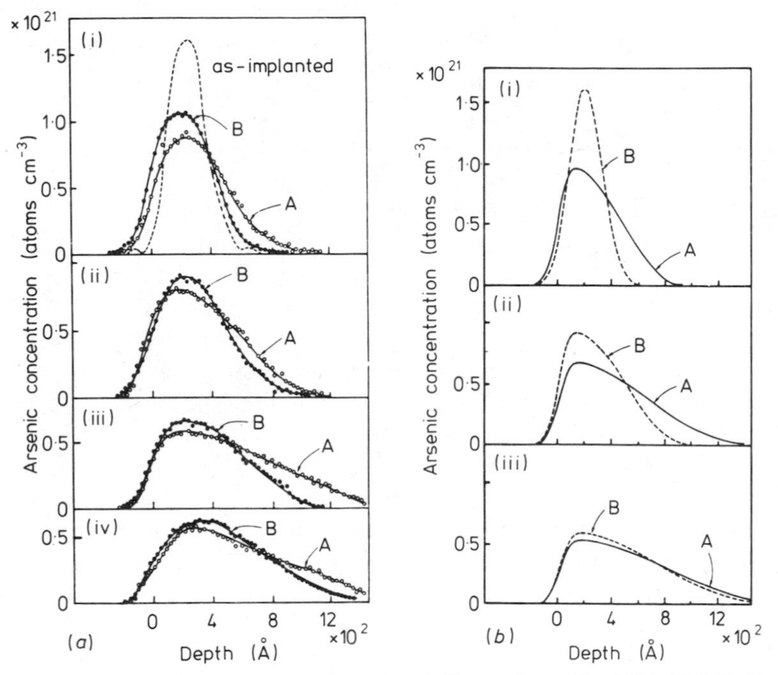

Figure 11. Experimental (*a*) and calculated (*b*) arsenic profiles (40 keV As^+) after ruby laser irradiation at several energy densities – in J cm^{-2} – (*a*): (i), 0·6; (ii). 1·05; (iii), 1·7; (iv), 2·2; – (*b*): (i), 0·75; (ii), 1·2; (iii), 1·7. The curves labelled A and B refer to 4500 Å and 500 Å thick amorphous layers respectively.

distribution is not symmetrical around the projected range R_p. The near-surface region acts as a non-permeable boundary and the lifetime of the liquid depends on the distance from the surface. The experiment pointed out that the broadening of the dopant is larger in the samples overlaid with the thicker amorphous layer. With increasing energy density the difference between the two profiles becomes smaller and smaller. These results reflect qualitatively the behaviour of the calculated spatial and temporal extent of the liquid layer. In the sample with the thicker amorphous layer the coupling with the laser energy is more effective and for the same energy density a larger amount of material is melted.

Taking into account the solid–liquid interface kinetics, as reported in figure 7, and solving numerically the diffusion equation the full curves shown in figure 11(*b*) were calculated. The differences between the calculated A and B profiles follow the same trend as the experiment. The main difference between the two profiles occurs for energy densities near the threshold value. For instance no broadening of the profile was computed in A below 0·7 J cm^{-2} energy density irradiation, this value being smaller than the threshold for surface melting of Si with a 500 Å amorphous layer.

In these previous experiments on B, P and As dopants a broadening of the profiles was detected after laser irradiation. The impurities in any case occupy substitutional lattice sites and quite reasonable values of the electrical parameters were also found. According to the melting hypothesis, the final impurity profile should be determined not only by the high diffusion coefficient in the liquid phase, but segregation effects might also play a role. For impurities like As, B and P the segregation coefficients do not differ significantly from 1; moreover the high solid–liquid interface velocity ($\sim 2\,m\,s^{-1}$) further decreases this effect.

Other experiments have been performed with impurities characterised by a low segregation coefficient. Consider for example the case of Cu, with an equilibrium segregation coefficient of 4×10^{-4}, low solid solubility and a very high diffusion coefficient in the solid phase. Cu ions were implanted at a depth of about 500 Å and then some samples were thermally annealed and others laser irradiated. The corresponding Cu profiles are reported in figures 12(*a*) and (*b*), together with the as-implanted profile for comparison. The measurements indicate that after $1\,J\,cm^{-2}$ ruby laser irradiation all the implanted Cu atoms accumulated at the sample surface. A different behaviour is found for thermally annealed samples. After 30 min at 500 °C the Cu profile becomes flat with a depth of about 700 Å. At annealing temperatures higher than 700 °C a reduction in the yield occurs. This reduction is in quantitative agreement with the data reported in the literature for the solid solubility.

The strong accumulation, caused by laser irradiation, of impurities at the sample surface observed in the case of Cu has also been seen for Pb, Te and Bi. Calculations for the distribution of impurities accounting for diffusion, time dependence of the melting and segregation coefficients by Baeri *et al* (1978) suggest that a fit of the experimental data requires a segregation coefficient larger than the equilibrium value.

In laser annealing one deals with a typical non-equilibrium process. The high velocity of the solid–liquid interface during freezing and the extremely elevated quenching rate could give rise to metastable phase formation. In addition the retained impurity concentration exceeds the solid solubility limit by several orders of magnitude. For almost all impurities in Si the phase diagram shows a retrograde solid solubility with a maximum at a temperature 200–300 °C below the melting point. In this case Jackson and Leamy

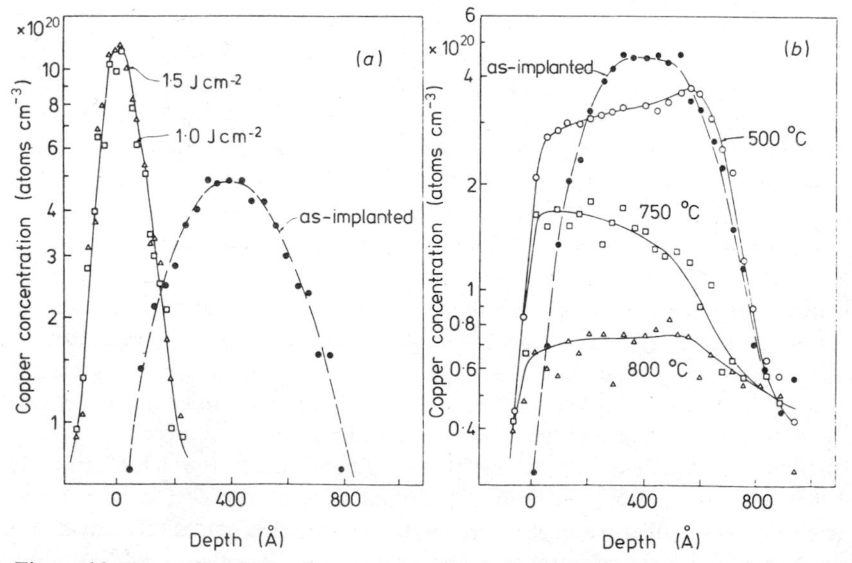

Figure 12. Cu profiles in Si after ruby laser irradiation of 1·0 and 1·5 J cm^{-2} (*a*) and after 30 min annealing at 500, 750 and 800 °C respectively (*b*). The as-implanted profile is shown for comparison.

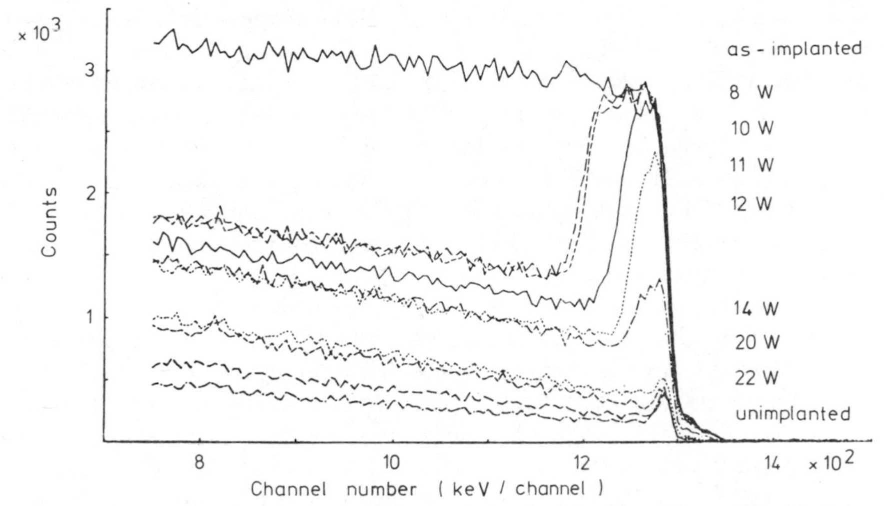

Figure 13. Aligned yields of a Si (100) sample implanted with self-ions at liquid nitrogen temperature after irradiation with CW laser at increasing powers.

(1979) pointed out that this limiting value should not be exceeded as it would be in a simple eutectic phase diagram. Impurity trapping at the moving solid–liquid interface must then be invoked.

3. CW irradiation

The experimental set-up for the use of a continuous mode laser requires a scanning system to achieve irradiation over a large wafer area. The beam spot is of the order of 10–100 μm in diameter while the area of interest is of the order of a few square centimetres. The advantage consists in the possibility of 'scribing' along a given pattern.

Two systems are commonly used: in one the beam moves over the sample and in the other the sample is driven mechanically by step motors. Several optical arrangements have been designed to move the beam. A prism and lens unit moves like a pendulum and the sample holder moves slowly in a perpendicular direction with a velocity of the order of 10^{-3} cm s^{-1} (Katchurin *et al* 1976). Another arrangement synchronously deflects the laser beam by two mirrors driven by galvanometers (Gat and Gibbons 1978). A focusing lens positioned in front of the first mirror controls the spot size on the sample. For this system the sample is located in the focal plane of the lens to avoid non-uniformity problems during the scan. After a scan along a line the beam is translated normally by 10–20 μm.

For a fixed optical line the sample is driven by step motors. Changes in the velocity occur at the start and at the end of the scan so that only the central region is uniformly illuminated. The scan velocity is of the order of a few centimetres per second.

Ion lasers such as Ar, Xe and Kr are usually employed with a power of about 10 W. The main difference between Q-switched and CW lasers consists in the duration: nanosecond for pulse annealing and millisecond for CW irradiation.

The difference in the time scale implies a different mechanism of annealing. The threshold behaviour associated with the amorphous to single crystal transition is replaced by a gradual decrease of the thickness of the disordered layer as shown in figure 13. The aligned yields recorded in Si samples overlaid with about 3000 Å thick amorphous layers for different powers of the laser indicate clearly that the regrowth takes place at the initial crystal–amorphous interface and proceeds to the sample surface (Rimini *et al* 1979). The spectra are similar to those obtained during thermal solid-phase regrowth of these samples. Thermal regrowth of amorphous Si occurs above 550 °C and requires several minutes. The laser-induced regrowth is accomplished in about 1 ms (Williams *et al* 1978). The thermal profile was calculated by Lax (1978) and it explains the experimental regrowth data (Gat *et al* 1979).

As in thermal regrowth the orientation of the underlying substrate plays an important role. Crystallisation induced by CW irradiation, on ⟨111⟩ oriented substrates is characterised by a large number of extended defects as pointed out by Auston *et al* (1978).

Another relevant characteristic of the continuous mode irradiation is the absence of any appreciable dopant profile broadening. As an example figure 14 shows the impurity profiles obtained by SIMS in arsenic-implanted Si substrates, under as-implanted, laser- and thermally annealed conditions. The laser-annealed profile is identical to the as-implanted one. No diffusion of the impurities has occurred during the laser annealing and the profile follows exactly the Pearson distribution (Gat *et al* 1978b). Similar results were obtained for B-implanted silicon.

The electrical activity is comparable to or better than that obtained by conventional thermal annealing; essentially all the impurities are electrically active. The crystallisation, as analysed by transmission electron microscopy, is almost perfect, with a small amount of residual disorder in the form of dislocation loops for the B-implanted Si (Gat *et al* 1978c).

Recently in Si, heavily doped with As, a substitutional concentration in excess of the solid solubility limit has been measured by Regalini *et al* (1979). The system is not stable

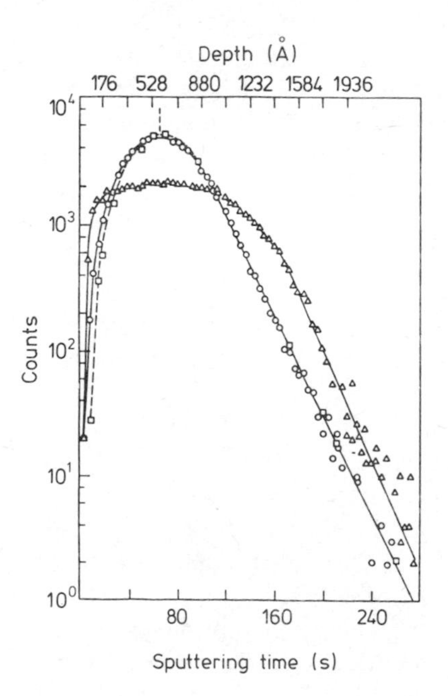

Figure 14. As concentration profile in As-implanted silicon after laser and thermal annealing. □, as-implanted; ○, laser annealed; △, thermally annealed (1000 °C, 30 min): ------- Pearson IV distribution with LSS range statistics.

and a subsequent thermal treatment induces a precipitation of As or at least an off-lattice displacement. The mechanisms involved like, for instance, As–defect or As–As pair formation are not yet detailed.

Irradiation with cw lasers has not been limited to ion-implanted semiconductors but has also been used for polycrystalline silicon layers. The major effect of the laser scan is to increase the grain size of the polycrystalline material. Electrical measurements indicate that essentially 100% electrical activation of the implanted dopants is obtained. Changes in the structure of the irradiated poly-layer are reported in figure 15 where the micrograph has been taken along the boundary of the line scan in cw annealing of a B-implanted Si sample. Within the scanned region there is clear evidence of crystal growth. Columnar crystallites extending from the laser scan boundaries are noted. Their shape is long and narrow and they are aligned parallel to the line scan, with a maximum length of about 32 μm and an average width of 2 μm. The initial structure consisted of a fine-grained structure with an average grain size of about 550 Å (Gat *et al* 1978a).

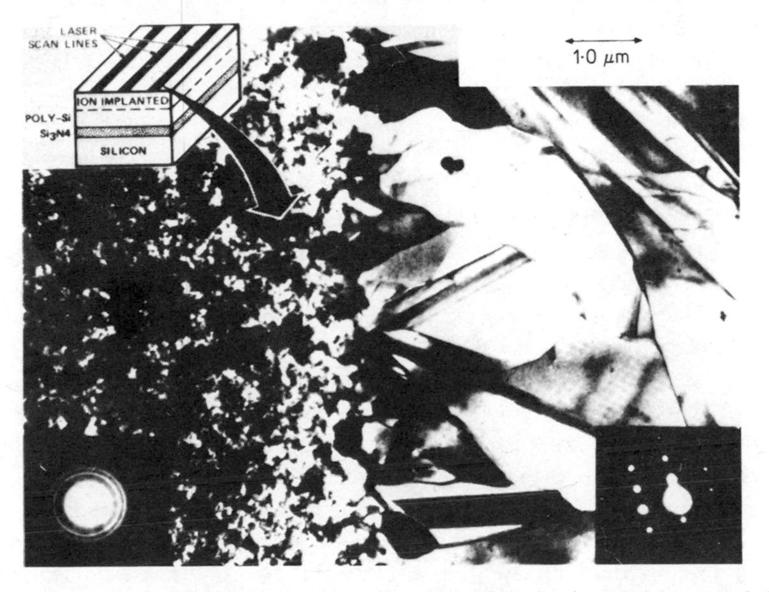

Figure 15. Transmission electron micrograph of ion-implanted, laser-annealed poly-Si film at boundary of laser beam line; selected diffraction patterns characteristic of each region are shown in the inset.

4. Contacts formation

The formation of Ohmic contacts is one of the crucial steps in semiconductor technology. The use of a metal–silicon compound, instead of a metal film, offers several advantages: silicides are stable under aging and low-temperature heat treatment. A new surface is created, the silicon near the interface being consumed in the reaction. Many of the impurities from the original silicon surface are redistributed at the outer silicide surface. Silicides also provide a good control of the Schottky barrier height and good adhesion to silicon (for a review see Tu and Mayer 1978).

Silicides are formed by heating a thin metal layer, about 1000 Å thick, deposited on the silicon substrates. Laser irradiation could also offer some advantages in this case over the conventional furnace approach: the processing time is shorter; a limited area of the device is heated; and for nanosecond or shorter laser pulses the process can be performed in air.

The amount of work published on contact formation in Si and GaAs is much less than that published for the substrates themselves. In view of the relevance of contacts in the device area it is worthwhile to summarise the studies already published. Several silicon–metal systems have been investigated, e.g., Pd–Si, Pt–Si, Mo–Si, Co–Si, with metal film thicknesses ranging from a few hundredths to a few thousand ångström. In almost all the cases, after irradiation with a Nd:YAG Q-switched laser, a mixed metal–silicon structure has been found. Many silicides have been detected by x-ray diffraction (Wittmer and von Allmen 1979, Poate *et al* 1978, 1979). The thickness of the reacted layer increases with the energy density of the light pulse and it seems to correlate with the melting point of the metal. The reacted layer is laterally uniform and its resistivity compares well with that of the silicides prepared by conventional heating.

It must be pointed out that because of its high reflectivity the metal layer absorbs only a small fraction, typically 10%, of the incident laser energy. Single shots of a few joules per square centimetre are required for the reaction.

The mechanisms involved in the silicide formation also seem to be associated with the melting of the overlying metal layer; the surface of the metal layer melts, the melt extends in depth to reach the silicon substrate. Intermixing with a coefficient of about $10^{-4} cm^2 s^{-1}$ takes place with silicon. The concentration gradient before freezing is also responsible for the phase formation, together with the velocity of the liquid–solid interface. Direct information on the melting has been provided by *in situ* measurements of time-resolved reflectivitiy in the Pd–Si system by von Allmen and Wittmer (1979). From their experimental results and heat flow calculations, they deduce that the overlying metal film melts with the underlying Si crystal. When the liquid–solid boundary reaches the region of the eutectic mixture, its velocity is strongly retarded by the decreased eutectic melting point and by the liberation of the heat of fusion of the silicides.

Observations of the silicide structures support the formation of phases by nucleation and growth from a supercooled liquid. A cellular structure has been detected in the Co–Si system after irradiation (van Gurp *et al* 1979). Two types of cells were present: the small ones with diameters of about 0·1 μm and the large ones of about 1 μm diameter. The first structure was ascribed to the constitutional supercooling of the melted silicon. Impurities are rejected from the advancing solid phase and the walls of the cell consist of a second phase richer in impurity. The same effect, on a larger scale, is found in bulk metals. The formation of the large cells, whose size is comparable with the liquid layer thickness, is attributed to convective motion resulting from pressure gradients.

Liquid formation does not necessarily imply that the metal film must be completely melted for the reaction to proceed in the liquid phase. Metal–Si eutectic temperatures (830 and 720 °C for Pt and Pd with Si respectively) are much lower than the metal melting temperatures. Melting can therefore start at the interface at relatively low temperatures. All these aspects deserve further investigation.

Silicide formation has also been accomplished by CW laser irradiation. In this case the experiments cannot be performed in air and a non-oxidising environment must be adopted. Uniform layers of PdSi and Pd_2Si have been obtained by scanning an Ar-laser beam (Liau *et al* 1979a) over a few thousand ångström thick Pd layer deposited on Si single crystal. Calculations on the basis of solid-phase thin-film reactions are in agreement with the experimental data as shown by Liau *et al* (1979b).

In addition to silicides, the formation of contacts on GaAs has been obtained by pulse annealing in high-dose Te implants with a Q-switched Nd:YAG laser and by removing the excess surface Ga (Barnes *et al* 1978). In other experiments Ohmic contacts of excellent electrical and morphological quality were obtained by deposition of Au–Ge followed by irradiation of suitable energy density (Eckhardt *et al* 1979, Gold *et al* 1979).

5. Conclusions

The processing of ion-implanted silicon either by rapid laser pulses or by scanning CW lasers produces crystalline layers of good quality. The effects of a high-power laser pulse can be interpreted on the basis of the occurrence of a liquid layer, as it involves both the transition to single crystal and the impurity redistribution. The results of CW-laser annealing can be predicted on the basis of processes occurring in the solid phase. The details of the temperature profile must be considered to account for the formation of metal–silicon mixtures.

References

von Allmen M and Wittmer M 1979 *Appl. Phys. Lett.* **34** 68

Auston D H, Golovchenko J A, Smith P R, Surko C M and Venkatesan T N C 1978 *Appl. Phys. Lett.* **33** 539

Baeri P, Campisano S U, Foti G and Rimini E 1978 *Phys. Rev. Lett.* **41** 1246

Barnes P A, Leamy H J, Poate J M, Ferris S D, Willams I S and Celler G K 1978 *Appl. Phys. Lett.* **33** 412

Bean J C, Leamy H J, Poate J M, Rozgonyi G A, Sheng T T, Williams J S and Celler G K 1978 *Appl. Phys. Lett.* **33** 227

Campisano S U, Catalano I, Foti G, Rimini E, Eisen F and Nicolet M A 1978 *Solid-St. Electron.* **1** 485

Csepregi L, Mayer J W and Sigmon T W 1976 *Appl. Phys. Lett.* **29** 92

Eckhardt G, Anderson C L, Hess L D and Krumm C F 1979 *Laser–Solid Interactions and Laser Processing: AIP Conf. Proc. No.* 50 p 641

Foti G, Della Mea G, Jannitti E and Majni G 1978d *Phys. Lett.* **68A** 368

Foti G, Rimini E, Bertolotti M and Vitali G 1978a *Thin Film Phenomena: Interface and Interactions* Ed I E Baglin and J M Poate (New York: The Electrochemical Society) p 88

—— 1978b *Phys. Lett.* **65A** 430

Foti G, Rimini E, Tseng W F and Mayer J W 1978c *Appl. Phys.* **15** 365

Gat A, Gerzberg L, Gibbons J F, Magee T J, Peng J and Hong J D 1978a *Appl. Phys. Lett.* **33** 775

Gat A and Gibbons J F 1978 *Appl. Phys. Lett.* **32** 142

Gat A, Gibbons J F, Magee T J, Peng J, Deline V R, Williams D and Evans Jr C A 1978b *Appl. Phys. Lett.* **32** 276

—— 1978c *Appl. Phys. Lett.* **33** 389

Gat A, Lietoila A and Gibbons J F 1979 *J. Appl. Phys.* **50** 2926

Gerasimenko N N, Dvurechesky A V, Katchurin G A, Pridachin N B and Smirnov L S 1975 *Proc. Int. Conf. Ion Implantation, Budapest* ed J Gyulai and E Pasztor (Budapest: Central Research Institute for Physics) p 263

Gold R B, Powell R A and Gibbons J F 1979 *Laser–Solid Interactions and Laser Processing: AIP Conf. Proc. No.* 50 p 635

Golovchenko J A and Venkatesan T N C 1978 *Appl. Phys. Lett.* **32** 147

van Gurp G J, Eggermont G E J, Tamminga Y, Stacy W T and Gijsbers I R M 1979 *Appl. Phys. Lett.* **35** 273

Gyulai J and Revesz P 1979 *Defects and Radiation Effects in Semiconductors 1978: Inst. Phys. Conf. Ser.* 46 p 128

Jackson K A and Leamy H T 1979 *Laser–Solid Interactions and Laser Processing: AIP Conf. Proc. No.* 50 p 102

Katchurin G A, Nidew E V, Khodjacsih A B and Kovaleva L A 1976 *Fiz. Tekh. Poluprov.* **10** 1890

Khaibullin I B, Titov V V, Shtyrkov E I, Zaripov M M, Stashko V P and Kuzmin K P 1975 *Proc. Int. Conf. Ion Implantation, Budapest* ed J Gyulai and E Pasztor (Budapest: Central Research Institute for Physics) p 212

Lax M 1978 *Appl. Phys. Lett.* **33** 786

Liau Z L, Tsaur B Y and Mayer J W 1979a *Laser–Solid Interactions and Laser Processing: AIP Conf. Proc. No.* 50 p 509

—— 1979b *Appl. Phys. Lett.* **34** 221

Mayer J W, Eriksson L and Davies J A 1970 *Ion Implantation in Semiconductors* (New York: Academic Press)

Poate J M, Leamy H J, Sheng T T and Celler G K 1978 *Appl. Phys. Lett.* **33** 918

—— 1979 *Laser–Solid Interactions and Laser Processing: AIP Conf. Proc. No.* 50 p 527

Regalini J L, Sigmon T W and Gibbons J F 1979 *Appl. Phys. Lett.* **35** 114

Revesz P, Farkas G, Mezey G and Gyulai J 1978 *Appl. Phys. Lett.* **33** 431

Rimini E, Baglin I E and Sedwick T 1979 unpublished

Shtyrkov E I, Khaibullin I B, Galyatudinov M G and Zaripov M M 1975a *Opt. Spektrosk.* **38** 1031

Shtyrkov E I, Khaibullin I B, Zaripov M M, Galyatudinov M G and Bayazitov R M 1975b *Fiz. Tekh. Poluprov.* 9 2000

Tseng W F, Mayer J W, Campisano S U, Foti G and Rimini E 1978 *Appl. Phys. Lett.* **32** 824

Tu K N and Mayer J W 1978 *Thin Films – Interdiffusion and Reactions* ed J M Poate, K N Tu and J W Mayer (New York: Wiley) ch. 10

Vitali G, Bertolotti M, Foti G and Rimini E 1978 *Appl. Phys.* **17** 111

White C W, Christie W H, Appleton B R, Wilson S R, Pronko P P and Magee C W 1978 *Appl. Phys. Lett.* **33** 662

Williams J S, Brown W L, Leamy H J, Poate J M, Rodgers J W, Rousseau D, Rozgonyi G A, Shelnutt J A and Sheng T T 1978 *Appl. Phys. Lett.* **33** 542

Wittmer M and von Allmen M 1979 *J. Appl. Phys.* **50** 4786

The role and effects of Cl in the thermal oxidation of silicon

G J Declerck†
Katholieke Universiteit Leuven, Departement Elektrotechniek, Laboratorium ESAT, Kardinaal Mercierlaan 94, 3030 Heverlee, Belgium

Abstract. This paper reviews the effects of chlorine-containing ambients during thermal oxidation of silicon on the properties of the silicon dioxide layer of the Si–SiO_2 interface and of the silicon bulk material.

Based on the assumption of thermodynamic equilibrium the reaction products at high temperatures of HCl/O_2-mixtures, of TCE/O_2 and C33/O_2 are compared in order to explain the differences observed between TCE oxides and HCl oxides. Furthermore C33 oxides are shown to be equivalent to HCl oxides.

The kinetics of oxide growth and the chlorine incorporation in the oxide are discussed. All analytical techniques show a peak of the chlorine distribution very close to the Si–SiO_2 interface. The existence of a chlorine-rich second phase has recently been pointed out.

The use of a chlorine-containing ambient strongly reduces the oxide defect density, even at temperatures as low as 900–1000°C. Chlorine oxidation is also very effective in decreasing the number of mobile ions introduced during oxidation and in cleaning the furnace tubes prior to oxidation. The sodium neutralisation capability and the interface state density of chlorine oxides will be discussed.

One of the most important advantages of the technique is the improvement of the minority carrier generation lifetime and the elimination of localised leakage current spots. This can be attributed to two possible mechanisms: first, the removal of volatile metallic chlorides, and second, the effect on silicon lattice defects where metallic impurities preferentially precipitate. The behaviour of oxidation-induced stacking faults during chlorine oxidation is discussed in detail. The effect of HCl on the oxygen content of the silicon is also pointed out.

1. Introduction

The performance of silicon integrated circuits and the fabrication yield of modern LSI processes depend for a great deal on the quality of the silicon thermal oxidation technology. During the last 15 years, considerable effort has been directed towards a better understanding of the physics and the electrical behaviour of thermally grown SiO_2. Improved oxidation techniques have been looked for.

This paper deals with the effects of a chlorine-containing ambient on the properties of the oxide layer, the oxide–silicon interface and the silicon bulk material. The thermodynamic reaction products in the oxidation furnace, resulting from the addition of different chlorine compounds will be compared. The kinetics of oxide growth and the chlorine incorporation in the oxide are discussed. The influence on oxide defect density

† Fellow (onderzoeksleider) of the Belgian National Research Foundation (NFWO).

0305-2346/80/0053-0133 $03.00

and on interface trap density and the neutralisation capability of mobile ionic charges are reviewed. Finally the improvement of minority carrier lifetime and the reduction of localised leakage currents will be discussed in connection with the effects on silicon crystalline defects as oxidation-induced stacking faults and with the removal or redistribution of metallic impurities.

During recent years several chlorine-containing compounds have been evaluated as possible candidates for improving the SiO_2 quality. The beneficial effect on the minority carrier lifetime caused by the addition of small amounts of HCl during oxidation was one of the first advantages indicated (Robinson and Heiman 1971, Ronen and Robinson 1972). At that time people were particularly interested in the 'gettering effect' of the HCl oxidation. Kriegler *et al* studied in great detail the properties of oxides grown in the presence of small amounts of HCl or Cl_2 (Kriegler *et al* 1972a, Kriegler 1972b, 1973a). These oxides will be referred to as HCl oxides or Cl_2 oxides. Kriegler put the emphasis on the sodium passivation capability of these oxides.

Chen and Hile (1972) have introduced the use of trichloroethene, C_2HCl_3 referred to as TCE, mainly because of its easier handling as compared to the highly corrosive HCl gas. This led to the TCE oxides, the basic properties of which have been described in the literature (Declerck *et al* 1975, Janssens and Declerck 1976, Hattori 1978a). Most recently the addition of 1,1,1-trichloroethane, $C_2H_3Cl_3$ referred to as C33, has been investigated (Janssens and Declerck 1978a, Linssen and Peek 1978). It has been shown that C33 oxides are equivalent to HCl oxides whereas TCE oxides have properties much more similar to those of Cl_2 oxides. Other chlorine compounds such as CCl_4 have also been mentioned (Young and Osburn 1973).

The introduction of these different techniques and the literature available on the various chlorine compounds give rise to a considerable confusion when the properties of chlorine oxides are discussed. Therefore in an attempt to clarify this, first of all some results of a thermodynamic analysis will be discussed. This analysis will give more insight into the relationship between HCl oxides, TCE oxides, Cl_2 oxides and C33 oxides.

2. Thermodynamic analysis

The HCl gas added to the oxidation ambient will react with the oxygen according to the following thermodynamic equilibrium:

$$4\,HCl + O_2 \rightleftharpoons 2\,H_2O + Cl_2 \qquad (1)$$

If other chlorine compounds such as TCE or C33 are used an overall reaction as expressed here for TCE, has to be considered

$$4\,C_2HCl_3 + 9\,O_2 \rightleftharpoons 2\,H_2O + 6\,Cl_2 + 8\,CO_2 \qquad (2a)$$

$$2\,H_2O + 2\,Cl_2 \rightleftharpoons 4\,HCl + O_2 \qquad (2b)$$

The long transit time (typically 1 min) of the reacting gases in the hot zone of the furnace validates the assumption made above of thermodynamic equilibrium. Based on this the gas-phase composition in the oxidation ambient can be calculated. It should be remarked that in this way no information is obtained about the concentration of HCl, Cl_2 or H_2O within the growing oxide. This would only be possible if the maximum solubility of these gases in the oxide and the diffusivity through the oxide could be accounted for; these data however are not available. But it should be emphasised that when evaluating different chlorine-containing additives, the comparison should be based on the equilibrium gas-phase composition present in the oxidation ambient rather than, e.g., on the total number of chlorine atoms present in the chlorine compound. There is no fundamental reason to expect the same oxidation rate enhancement from 2·5% Cl_2 as from 5% HCl. Such reasoning leads to erroneous conclusions as will be demonstrated later.

A thermodynamic analysis of the gas composition in the oxidation ambient has been given first for HCl by van der Meulen and Cahill (1974) in their study of the oxidation kinetics of HCl oxides. A more extensive calculation was made by Tressler *et al* (1977), who correlated the equilibrium partial pressures of the reaction products in the Cl–H–O system to the oxide properties. They also suggested that the growth kinetics and the

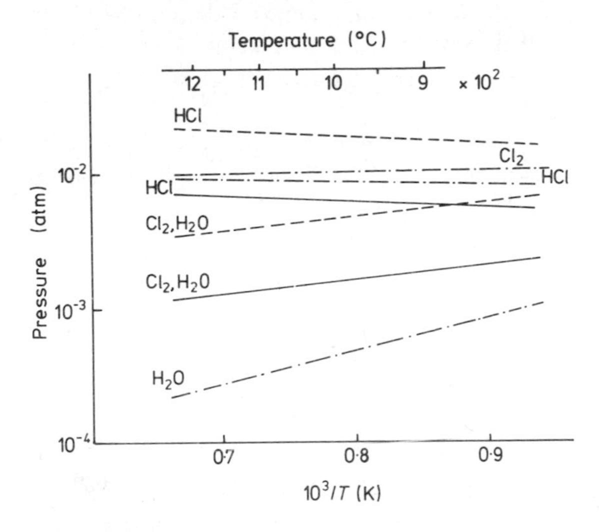

Figure 1. Equilibrium pressure of HCl, Cl_2 and H_2O as a function of the oxidation temperature. The additives are respectively 1% HCl (——), 1% C33 (-------) and 1% TCE (–·–) (Janssens and Declerck 1978a).

electrical characteristics of HCl oxides are related to the concentration of Cl_2 or ClO in the gas phase and not to the total chlorine content of the gas phase.

Janssens and Declerck (1978a) made a thermodynamic study for additions of HCl, TCE and C33. The gaseous reaction products as a function of temperature are shown in figure 1 for, resepectively, 1% HCl, 1% TCE or 1% C33. Since the concentrations of the reaction products are directly proportional to the additive concentration as long as lean mixtures are used (additive to O_2 ratio smaller than 10%), these figures can also be used for other additive concentrations. From that study, the following conclusions can be drawn.

(i) At temperatures above 600 °C the free energy of formation of HCl is lower than that of H_2O. As a consequence, at normal oxidation temperatures the hydrogen reacts to form HCl, leaving much smaller amounts of H_2O and Cl_2. This has also been pointed out earlier by other groups (van der Meulen and Cahill 1974, Tressler *et al* 1977).

(ii) The use of TCE leads to an oxidation atmosphere containing almost equal amounts of Cl_2 and HCl. This explains the Cl_2-like behaviour of TCE.

(iii) The addition of 1% C33 gives an oxidation ambient similar to the addition of 3% HCl. The physical and electrical properties of C33 oxides can be compared to those of HCl oxides if the 1 to 3 relationship between the concentrations is taken into account. For a discussion of secondary reaction products such as CO_2 or $COCl_2$ we refer to the literature (Janssens and Declerck 1978a).

3. Oxidation kinetics

The thermal oxidation of silicon described by the classical combination of a linear reaction-rate-controlled regime and a parabolic diffusion-controlled regime can be written as (Deal and Grove 1965)

$$x_o^2 + Ax_o = B(t + \tau) \tag{3}$$

where x_o is the oxide thickness, t is the oxidation time, B is the parabolic rate constant, B/A is the linear rate constant and τ accounts for the initial rapid oxidation phase. The parameters A and B have been studied for different oxidation ambients (O_2 or H_2O), as a function of substrate doping and orientation, and for various temperatures and pressures.

The use of chlorine compounds gives rise to a marked increase in the oxide thickness for the same oxidation time. This is illustrated in figure 2 where x_o is plotted against

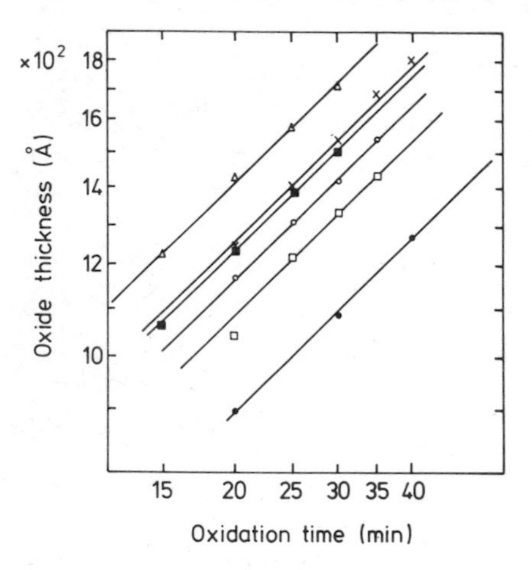

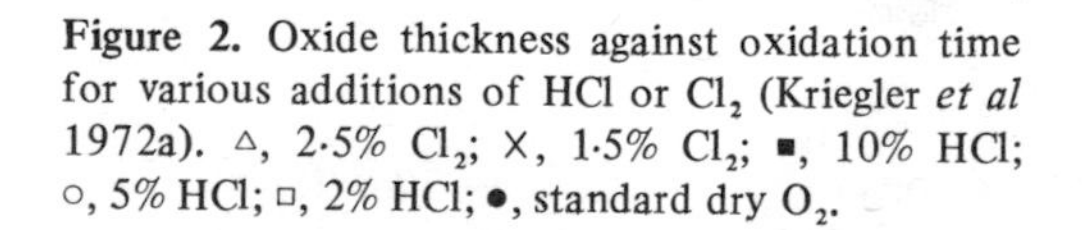
Figure 2. Oxide thickness against oxidation time for various additions of HCl or Cl_2 (Kriegler *et al* 1972a). △, 2·5% Cl_2; ×, 1·5% Cl_2; ■, 10% HCl; ○, 5% HCl; □, 2% HCl; ●, standard dry O_2.

oxidation time in the parabolic regime for different additions of Cl_2 or HCl (Kriegler *et al* 1972a). This figure shows that both HCl and Cl_2 strongly enhance the oxidation rate. It should be remarked however that, as has been pointed out in the previous section, no similarity exists between 2·5% Cl_2 and 5·0% HCl. From this observation it is clear that the number of chlorine atoms is not the dominant factor in controlling the oxidation rate. Although several detailed investigations have been undertaken to clarify the basic mechanisms behind the oxidation rate enhancement, no unambiguous picture has been given so far (van der Meulen and Cahill 1974, Hirabayashi and Iwamura 1973, Deal 1978, Deal *et al* 1978a, Ritchey *et al* 1976, Singh and Balk 1978).

Tressler *et al* (1977) have rationalised the experimental results of Kriegler *et al* (1972a) and of Hirabayashi and Iwamura (1973) by plotting the parabolic rate constant B as a function of the equilibrium partial pressure of Cl_2 in the oxidising ambient. Figure 3 shows the increase of the rate constant B with the equilibrium chlorine concentration for both the HCl–O_2 and Cl_2–O_2 systems. This is particularly important for the latter

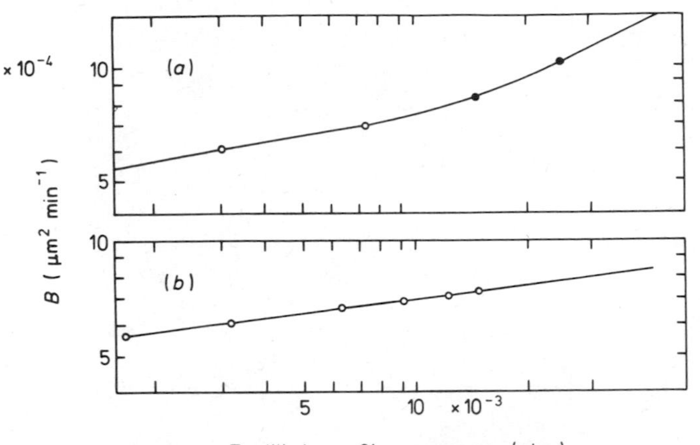

Figure 3. Enhancement of the parabolic growth rate, *B*, as a function of the equilibrium Cl_2 pressure at (*a*) 1150 °C and (*b*) 1100 °C. ○, $HCl–O_2$ oxidation; •, $Cl_2–O_2$ oxidation (Tressler *et al* 1977).

as in that case no H_2O can be present in the oxidising ambient. It proves that the diffusion rate of the oxidising species is strongly enhanced, probably because of the incorporation of chlorine in the growing oxide. The possible effect of the H_2O present in the $HCl–O_2$ system has been studied by Deal (1978) and by Deal *et al* (1978a). Singh and Balk (1978) compared TCE oxides with CCl_4 oxides and came to the conclusion that at lower temperatures (900 and 1000 °C), and for the same concentration of Cl_2 in the equilibrium gas phase, the parabolic oxidation rate of the TCE oxides is slightly higher in comparison with the CCl_4 oxides. At 1100 °C the rates are the same in both systems, indicating that under these conditions the oxidation kinetics are determined by O_2 and Cl_2 only.

The linear rate constant *B*/*A* which is mainly governed by the reaction mechanism itself is also increased by the chlorine oxidation. This is explained by the catalytic action of the chlorine present at the oxide–silicon interface. The first step of the overall reaction is the formation of a Si–Cl bond; this chlorination of the silicon is much faster than the oxidation itself and the reaction velocity is related to the etching of silicon in HCl or Cl_2 mixtures. The second step is the conversion of the silicon chlorides into silicon oxychlorides or SiO_2 because of the higher thermodynamic stability of the latter compounds. The two-step reaction may proceed faster than a single oxidation reaction (Kriegler 1973a, Hirabayashi and Iwamura 1973).

Recently Monkowski investigated in very great detail the physics and chemistry of the chlorine oxidation by means of various analytical techniques such as TEM, STEM, SEM and infrared spectroscopy (Monkowski 1978, Monkowski *et al* 1978). The most significant result of his work was the existence of an additional phase containing most of the chlorine and located at the oxide–silicon interface. This phase appears after a few minutes of oxidation as a great number of small regions, several hundred ånströms in diameter, and homogeneously distributed over the entire interface. As the oxidation continues, these regions grow and coalesce. Monkowski also proves that at the very moment of complete coalescence, complete sodium passivation is obtained. Further oxidation in a chlorine-containing ambient leads to a roughening of the oxide surface and

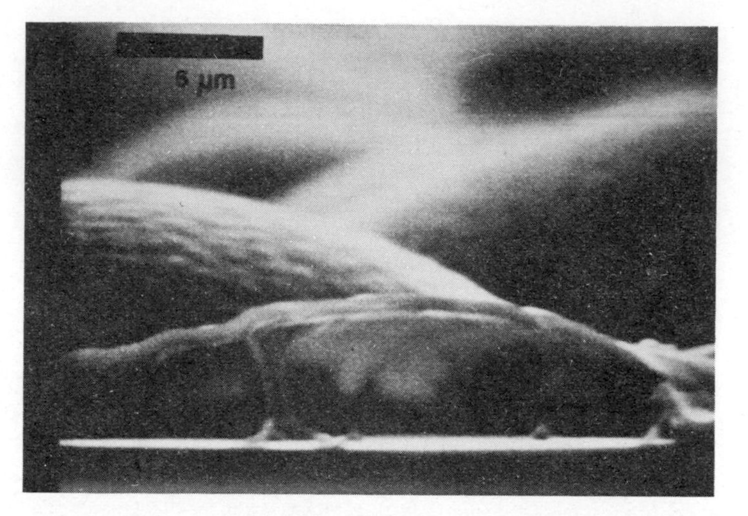

Figure 4. Cross-sectional SEM micrograph of oxide grown in 10% HCl at 1200 °C for 6 hours. The bubbles at the oxide–silicon interface and the additional condensed phase are clearly visible (Monkowski *et al* 1978).

later to the creation of bubbles (figure 4). The latter is due to the formation of a chlorine-rich gas lifting the oxide layer. Monkowski explains the growth of the additional phase and the formation of the interfacial gas by the thermodynamic equilibria at the silicon–oxide interface where two reactants, oxygen and chlorine, are present. The activity gradients of chlorine and oxygen are supposed to be different in the growing oxide and as a consequence the thermodynamic equilibrium at the interface changes as the oxidation progresses, allowing the formation under the oxide of additional chlorine-rich phases which would not be stable in high partial pressures of oxygen. The physical and electrical properties of chlorinated oxides are explained in terms of the presence of the additional chlorine-rich phase.

4. Chlorine distribution in the oxide

The distribution of the chlorine atoms within the chlorinated oxide has been studied by means of a variety of analytical techniques such as secondary ion mass spectroscopy (SIMS) (Kriegler *et al* 1973b, Frenzel *et al* 1979, Deal *et al* 1978b, Baxter 1973), Auger spectroscopy (Kriegler 1973a, Chou *et al* 1973), x-ray fluorescence (Monkowski 1978, Monkowski *et al* 1978, van der Meulen *et al* 1975), electron microprobe analysis (van der Meulen *et al* 1975) and nuclear backscattering (NBS) (Meek 1973, van der Meulen *et al* 1975, Butler *et al* 1977). Most of these techniques reveal a peaked distribution of the chlorine piling up within approximately 200 Å of the silicon interface. This is illustrated in figure 5 which shows a typical result obtained by SIMS (Deal *et al* 1978b). The peak concentration of the chlorine can be as high as $2 \times 10^{21}\ \mathrm{cm}^{-3}$ (Deal *et al* 1978b), whereas the concentration per unit area, averaged over the total oxide thickness, is in the range of 10^{15}–$10^{16}\ \mathrm{cm}^{-2}$ (Butler *et al* 1977). It should also be noted that the location of most of the chlorine in a thin layer near the interface agrees very well with the existence of the additional chlorine-rich phase proposed by Monkowski.

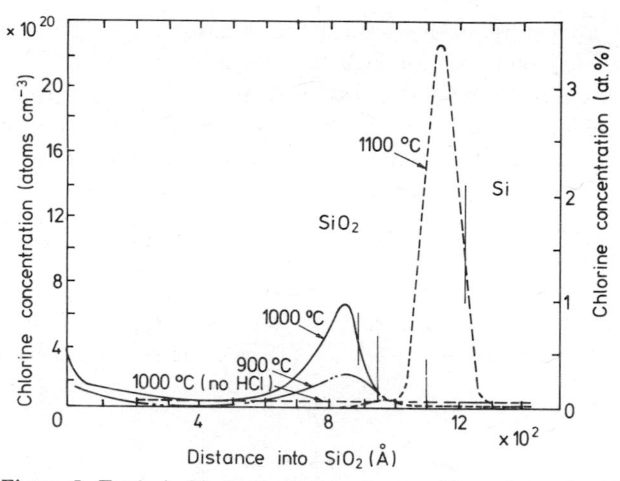

Figure 5. Typical chlorine concentration profile as determined by SIMS. Oxides were prepared in a 5% HCl/O_2 ambient at 900 °C, 1000 °C and 1100 °C. The $Si-SiO_2$ interfaces are indicated by solid vertical lines (Deal *et al* 1978b).

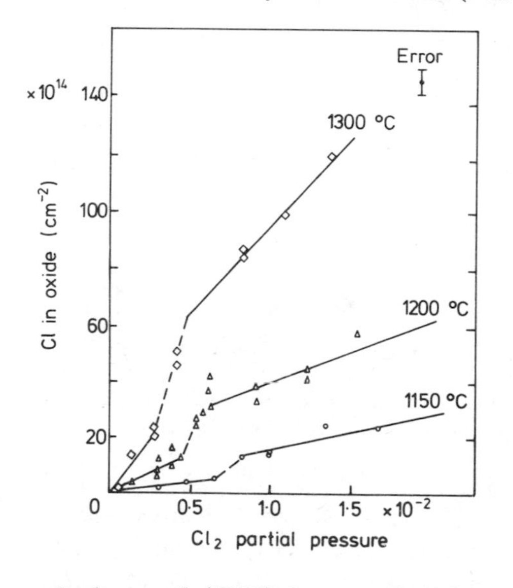

Figure 6. Chlorine content of the oxide as a function of the HCl concentration for various oxidation temperatures (Butler *et al* 1977, Monkowski 1978). Oxidation time, 30 min.

Butler *et al* (1977) have studied the total chlorine content of various oxides as a function of oxidation time and temperature, and HCl concentration. Figure 6 shows the chlorine content as a function of HCl concentration for various oxidation temperatures and for a fixed oxidation time. The break in each curve corresponds with the onset of second-phase formation (Monkowski 1978).

5. Properties of the oxide

5.1. Dielectric breakdown and oxide defect density

The oxide defect density, being the number per unit area of spots with low breakdown strength is one of the main yield limiting factors of modern LSI processing. The occurrence of oxide defects can be studied by making a statistical analysis of breakdown events as a function of applied voltage and capacitor area (Osburn and Ormond 1972a, 1972b). Typical breakdown distribution curves for a dry thermal oxide (DTO), a 1% C33 oxide and a 1% TCE oxide, grown at 1000 °C for 120 min, are shown in figure 7. These distributions are measured by an automated system as described by Osburn and Ormond (1972a). The thickness of the Al layer was only 30 nm to allow the self-healing breakdown to be distinguished from the final breakdown. It is clear that the distribution curves shift to higher breakdown voltages for the C33 and TCE oxides, although the maximum breakdown field is not influenced by the chlorine addition.

A detailed study of the breakdown properties of oxides prepared in different halogen-containing ambients has been carried out by Osburn (1974). The defect density of oxides

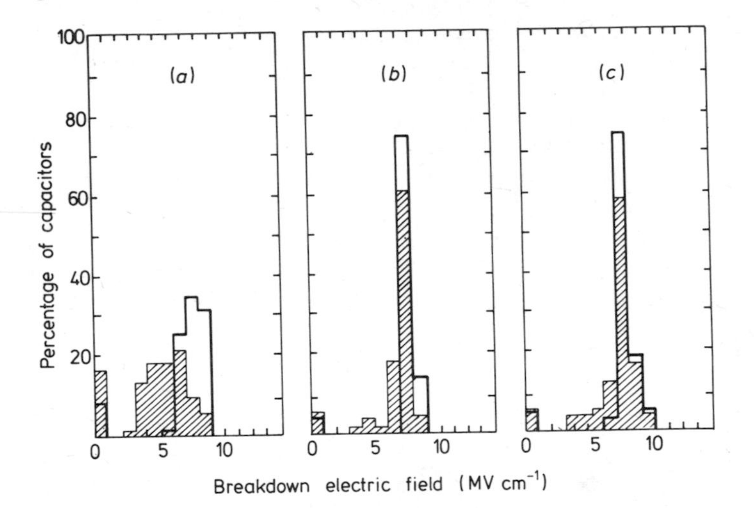

Figure 7. The distribution of the first self-healing (shaded area) and final breakdown electric field for: (*a*) a dry thermal oxide (DTO); (*b*) a 1% C33 oxide; and (*c*) a 1% TCE oxide; all grown at 1000 °C for 2 h. The capacitor area is 0·8 mm^2 and the calculated defect densities are, respectively, 120 cm^{-2}, 9 cm^{-2} and 15 cm^{-2} (Janssens and Declerck 1978a).

grown at 1000 °C to a thickness of 45 nm and using respectively HCl, CCl_4, C_2HCl_3, HBr or Cl_2 as additive are represented in figure 8. The defect density which is around 40 cm^{-2} for 0% halogen addition drops to a much lower value as soon as small amounts of halogen are added to the oxidising atmosphere. A minimum value is reached for 3%HCl addition. It is also seen that after an initial decrease the defect density increases drastically for Cl_2-additions greater than 2%. A similar effect is observed for TCE oxides, pointing again to the Cl_2-like behaviour of TCE.

The reduction of oxide defects is attributed to the removal of both electrical defects, as charged impurities or interface traps, and physical defects, as particulate inclusions and substrate bumps (Osburn 1974). It should be noted that the defect density improvement is observed at oxidation temperatures as low as 850 °C as illustrated in figure 9 (Osburn 1974). This might give the process engineer a very valuable argument to use HCl oxides even at these low temperatures where the sodium passivation effect and the stacking fault elimination can certainly not be attained, as will be discussed later. The rise of the defect density at higher chlorine concentrations is ascribed to the corrosive effect of the chlorine

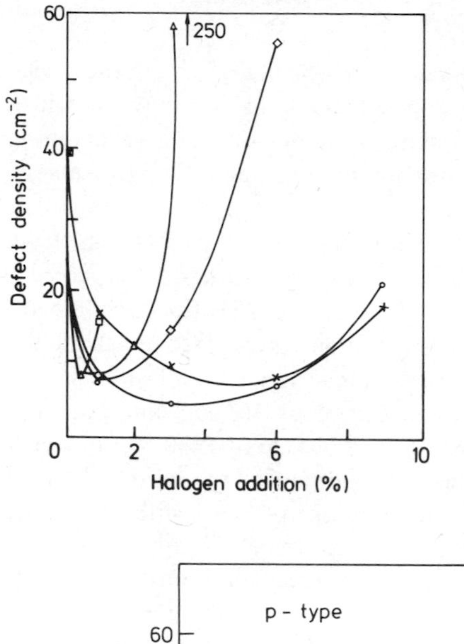

Figure 8. Effect of halogen addition during oxidation on defect density of SiO_2 films (Osburn 1974). ○, HCl; □, CCl_4; ◇, C_2HCl_3; ×, HBr; △, Cl_2.

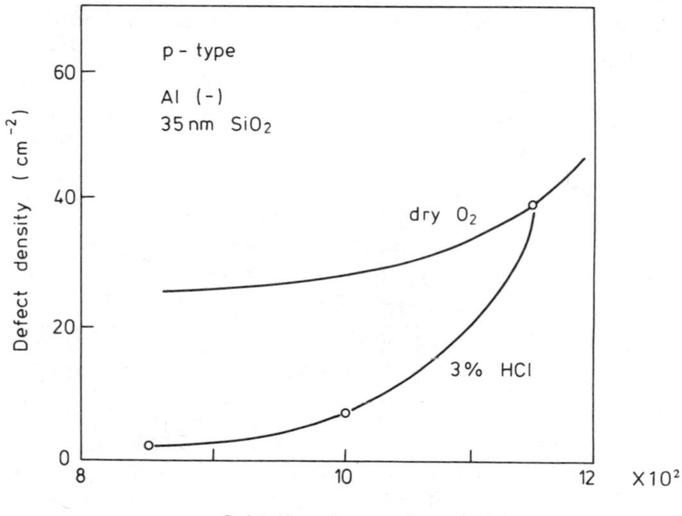

Figure 9. Dependence of defect density on oxidation temperature for dry and HCl oxide (Osburn 1974).

etching the silicon surface (Osburn 1974). At these chlorine concentrations non-uniform films and rough oxide surfaces are observed, indicating the presence of the chlorine-rich second phase.

Accelerated breakdown experiments have revealed reduced failure times for Cl_2 and CCl_4, whereas HCl, HBr, TCE and H_2O oxides have shown strongly improved wear-out reliability (Osburn 1974). It is believed that this effect is due to the presence in the oxidation atmosphere of hydrogen rather than halogens, since an improvement was also observed in oxides grown in H_2O–O_2 ambients or annealed at high temperatures in hydrogen (Osburn 1974).

5.2. Oxide charge, interface traps and oxide stability

The presence of mobile sodium ions in SiO_2 layers has a detrimental effect on the device stability. It causes threshold shifts, surface inversion, excessive leakage currents and a general degradation of the device performance. Sodium contamination can occur during the thermal oxidation itself but also during later process steps like annealing treatments, photolithography, etching or metallisation.

It was very soon observed that chlorinated oxides exhibited a passivation effect resulting in an improved electrical stability of the oxide (Kriegler *et al* 1972a, Kriegler 1972b, 1972c, 1973a, Kriegler *et al* 1973b, van der Meulen *et al* 1975). The phenomenon has been extensively investigated by Kriegler who demonstrated the threshold-like behavior of the passivation efficiency. The effective mobile ion concentration in Na-contaminated HCl oxides is plotted in figure 10 as a function of HCl concentration. The contamination was achieved by evaporating NaOH onto the oxide surfaces before metallisation. The mobile ion density was calculated from the oxide thickness and the observed flat-band voltage shift after a five minute positive B–T stress at 250 °C with an electric field of $5 \times 10^5 \, V \, cm^{-1}$. Radiotracer analysis, performed after the B–T stress showed the sodium to be located within a few hundred ångströms from the silicon interface (Kriegler *et al* 1973b). It is clear from figure 10 that very little passivation is obtained at 3% HCl, whereas at 4% and higher, complete passivation is reached. A similar threshold-like behaviour has been found as a function of HCl oxidation time. This is illustrated in figure 11 showing passivation efficiency against time for fixed HCl concentration and oxidation temperature. For TCE oxides and C33 oxides a threshold-like passivation behaviour has also been reported (Declerck *et al* 1975, Janssens and Declerck 1978a).

The mechanism of sodium passivation is not yet well understood. Quasistatic current–voltage measurements have shown that the sodium ions still migrate through the passivated oxides when an electric field is applied (Kriegler 1972b, Kriegler and Devenyi 1973, Janssens and Declerck 1978a). However, as soon as they reach the vicinity of the silicon interface they are trapped and their charge is neutralised, probably in the interaction with

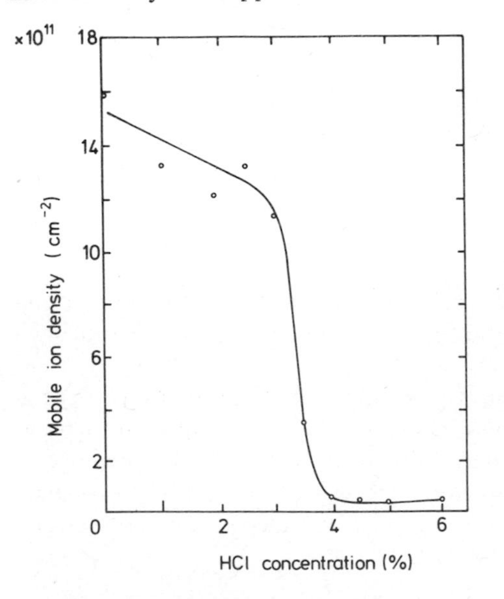

Figure 10. The effective mobile ion concentration in Na-contaminated HCl oxides as a function of HCl concentration. The oxides were grown at 1150 °C for 34 min (Kriegler *et al* 1973b).

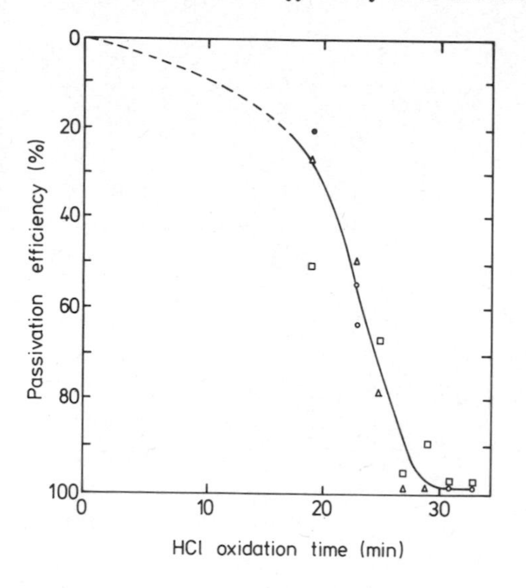

Figure 11. The passivation efficiency of HCl oxides as a function of HCl oxidation time. Different levels of Na contamination are presented as follows: ○, 4×10^{11} ions cm^{-2}; △, $1{\cdot}5 \times 10^{12}$ ions cm^{-2}; □, 4×10^{12} ions cm^{-2} (Kriegler *et al* 1973b).

the chlorine-rich layer close to the interface. The trapping efficiency has recently been studied by Stagg and Boudry (1978). Monkowski has clearly demonstrated that the coalescence of the second chlorine-rich phase and the sodium passivation reach completion at approximately the same conditions of oxidation temperature, time and HCl addition. He suggests that trapping and neutralisation of the sodium occurs only in these interfacial areas which contain the chlorine-rich phase (Monkowski 1978). The passivation efficiency is then proportional to the area covered by this phase. The relationship between passivation efficiency and chlorine content of the oxide, measured by means of α particle backscattering, has also been studied by Rohatgi *et al* (1977). They came to the conclusion that passivation is an increasing function of oxide chlorine content and is only weakly dependent on the level of sodium contamination over the range 5×10^{11} to 1×10^{13} cm^{-2}.

Large amounts of sodium ions, up to 10^{13} cm^{-2}, can effectively be stabilised in HCl oxides. However in modern LSI processing it is not recommended that this passivation effect should be relied upon as, at those oxidation conditions needed to give sufficient passivation efficiency, other electrical parameters of the oxide layer, such as the oxide defect density, are getting worse. It will be noted in the next section that the interface trap density will also rise by almost two orders of magnitude.

A much better approach is to reduce the total sodium contamination as much as possible throughout the complete processing and here HCl oxidation can still play an important role. Indeed it has been demonstrated that small amounts of chlorine significantly reduce the sodium content of thermally grown oxides. Furthermore, precleaning the quartz tube of the oxidation furnace by prolonged exposure to an O_2/Cl_2 mixture allows the fabrication of very clean oxides even if the oxidation itself is carried out without further addition of any chlorine-containing compound. In this way oxides having a sodium content smaller than 2×10^9 cm^{-2} have been obtained (Janssens and Declerck 1978a). It should be noted that the furnace-cleaning step should be of sufficient length to allow for the initially enhanced outdiffusion of sodium from the furnace tubes which has been observed (Kriegler 1973a, Janssens and Declerck 1978a, Mayo and Evans 1977).

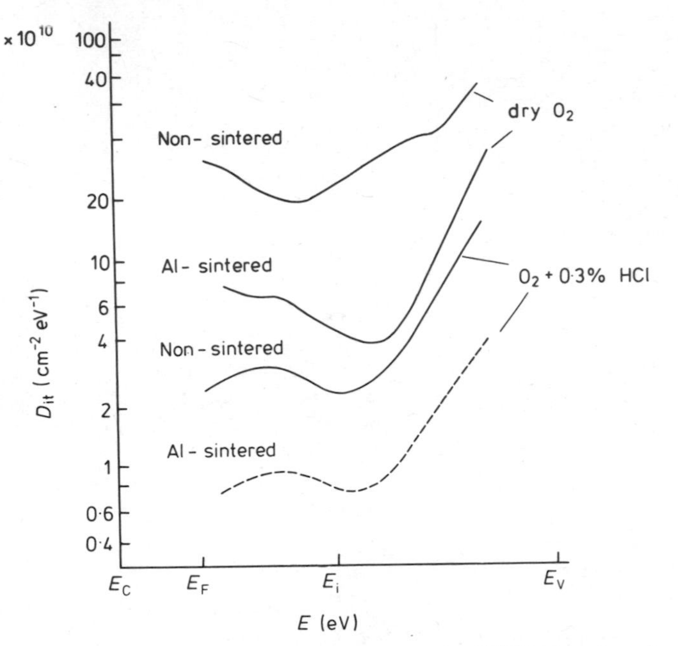

Figure 12. Interface trap density, D_{it}, at the Si/SiO_2 interface for ⟨100⟩-oriented silicon oxidised in dry O_2 and in O_2 + 0·3% HCl at 1100°C for 138 min (Severi and Soncini 1972). 1972).

Interface traps are generally believed to be caused by extrinsic defects such as chemical impurities and charged centres at the interface, or by structural defects such as dangling Si bonds (Goetzberger *et al* 1976). The density D_{it} of interface traps is directly related to 1/f-noise in MOSFETS, to charge loss in surface channel CCDS and to enhanced surface generation in both bipolar and MOS circuits. Values in the range 10^9–10^{10} $cm^{-2}eV^{-1}$ for ⟨100⟩-oriented silicon are typical of a good MOS process.

A strong reduction of interface trap density as compared with standard dry oxides has been reported for HCl oxides (Severi and Soncini 1972, Fogels and Salama 1971, Baccarani *et al* 1973, Singh *et al* 1974), TCE oxides (Declerck *et al* 1975) and C33 oxides (Janssens and Declerck 1978a). Figure 12 shows D_{it} for ⟨100⟩ material, oxidised at 1100°C in dry O_2 or in O_2 + 0·3%HCl. These distributions have been measured by means of the quasistatic $C-V$ technique (Kuhn 1970), the accuracy of which is about 1×10^{10} $cm^{-2}eV^{-1}$. The Al-sintering refers to a 525 °C annealing for 10 min in nitrogen after aluminium deposition; this step is known to strongly decrease the interface trap density (Balk 1965). The low D_{it}-values, lying at the limit of sensitivity of this measuring technique, have been attributed to the presence in the oxidising ambient of active hydrogen or hydroxyl groups because of the decomposition of HCl, TCE or C33 (Severi and Soncini 1972, Singh and Balk 1978). It is suspected that these hydrogen and/or hydroxyl groups are very effective in saturating the free silicon bonds at the interface (Kooi 1965, Montillo and Balk 1971). It is obvious however that the gettering or cleaning effect of the chlorine which is believed to remove sodium and other metal impurities from the oxidising environment may certainly not be ruled out. Indeed the chlorine oxidation technique provides a clean atmosphere during oxidation, minimising in this way the possibility of contamination of the oxides by sodium or other impurities which can give rise to higher interface trap densities.

It has to be emphasised again that as soon as complete sodium passivation is reached the interface trap density rapidly increases. This is illustrated in figure 13 showing D_{it} as a function of C33 concentration (Janssens and Declerck 1978a). It is seen that D_{it} increases from a low $10^9\,cm^{-2}\,eV^{-1}$, as measured by quasistatic capacitance–voltage technique and confirmed by more sensitive transfer loss measurements on surface channel CCDs, to $10^{11}\,cm^{-2}\,eV^{-1}$. This effect is ascribed to the excessive corrosion of the silicon surface by the chlorine.

It is generally accepted that the addition of small amounts of chlorine compounds does not have a pronounced effect on the fixed oxide charge (Kriegler *et al* 1972a,

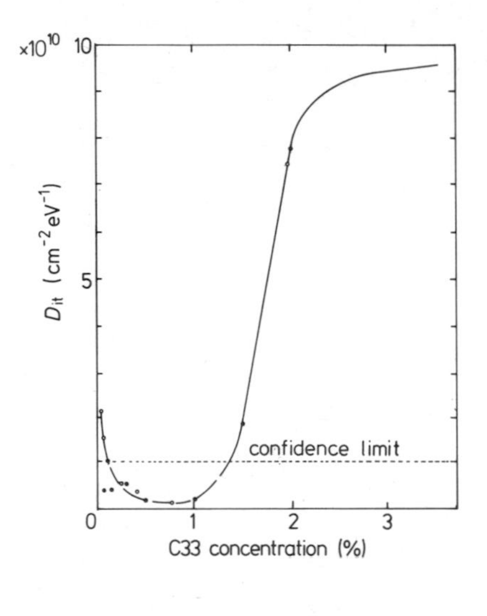

Figure 13. The interface trap density, D_{it}, as a function of C33 addition. The passivation threshold is lying at about 1% C33 (Janssens and Declerck 1978a). •, 1150 °C, 32 min; ○, 1200 °C, 30 min.

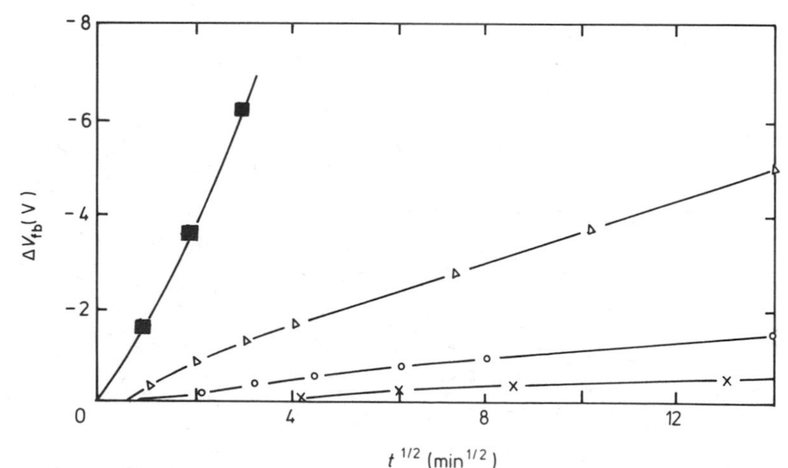

Figure 14. Change of flat-band voltage with time under negative bias in HCl oxide (100 nm, 9%HCl, 1150°C, Al gate, negative bias). Voltage applied: ■, 80 V; △, 75 V; ○, 70 V; ×, 60 V (van der Meulen *et al* 1975).

Janssens and Declerck 1976, Severi and Soncini 1972). Only in those cases where the standard dry oxides are not subjected to a proper postoxidation annealing, has a reduction of fixed oxide charge been observed (Janssens and Declerck 1978a).

Under application of high electric fields (>6 MV cm^{-1}) room temperature instabilities have been measured for oxides containing high amounts of chlorine (van der Meulen *et al* 1975, Heald *et al* 1976). Figure 14 shows the flat-band voltage shift of a 100 nm thick oxide grown at 1150 °C in O_2+9% HCl. The origin of these instabilities is not yet known. However it should be emphasised that under normal operating conditions oxides containing small amounts of chlorine do not show this kind of instability.

6. Properties of the silicon bulk material

The most striking advantage of chlorine oxides is probably the improvement of minority carrier lifetime and the associated reduction of leakage current. This phenomenon is usually studied by analysing the thermal relaxation curve of MOS capacitors pulsed from flat-band or inversion into deep depletion (Zerbst 1966, Schroder and Nathanson 1970, Schroder and Guldberg 1971). If generation within the depletion layer is dominant, the storage time, which is the time necessary to reach thermal equilibrium, is a good measure for characterising the generation lifetime.

The storage time improvement has first been reported by Robinson and Heiman (1971). They observed that a 16 h gettering step at 1200 °C in O_2+1% HCl resulted in a dramatic increase of the minority carrier lifetime from 0·3 μs for a standard steam oxidised wafer to 100–300 μs for the HCl-treated samples. A one hour oxidation at 1100 °C in 1% HCl/O_2 was found to give a minority carrier lifetime of 10–40 μs. Ronen and Robinson (1972) reported similar results for Cl_2 and HCl gettering at temperatures between 800 and 1150 °C although the lifetimes they found were considerably lower (4 μs for HCl oxidation at 1150 °C). They also observed that the lifetime improvement was much smaller in wafers having a higher O_2 content. This effect is particularly interesting with respect to the role of oxygen in the nucleation of oxidation-induced stacking faults (Mahajan *et al* 1977, Rozgonyi *et al* 1976).

The minority carrier generation rate in MOS capacitors has been studied very extensively by Young and Osburn for additions of HCl, Cl_2, CCl_4 or TCE (Young and Osburn 1973). For each of these additions a very significant improvement was found, as illustrated in figure 15, where the statistical distribution of storage times is drawn for various oxides. A dramatic improvement of over two to three orders of magnitude is seen for the Cl_2 and HCl oxides as compared with the control oxides. The use of N_2/HCl anneals before or after dry oxidation did not result in any storage time improvement, but gave oxides which were even worse than the control oxides. Young and Osburn also point out that after stripping the chlorine oxide, a second oxidation in a chlorine-free atmosphere still results in a high storage time. This indicates that indeed some of the impurities or defects which are believed to be responsible for the low storage times in standard oxides are effectively removed by prior chlorine oxidation.

Most of the early papers on the lifetime improvement due to chlorine oxidation attribute the effect to a gettering of heavy metals through the formation of volatile chlorides (Robinson and Heiman 1971, Ronen and Robinson 1972, Young and Osburn 1973). Green *et al* (1974) used Auger analysis to study the concentration of metals such

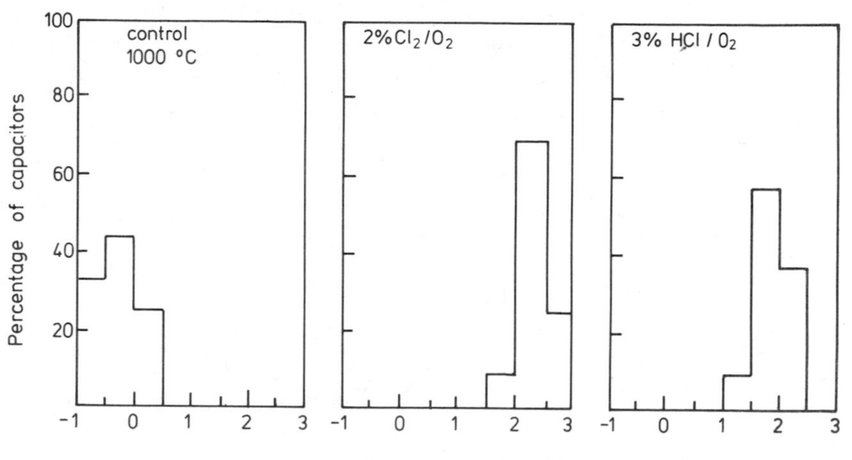

Figure 15. Statistical distributions of inversion times for dry O_2, Cl_2 and HCl oxides (Young and Osburn 1973).

as Cu, Fe or Au in silicon after different high temperature or HCl treatments. They found out that the removal of these metals was much smaller than would be predicted from diffusion-rate limited processes. They suggested that the heavy metals may be associated with the presence of crystalline defects and that the rate limiting process may be the release of the metals from the defect or from the precipitate.

It is clear that one of the mechanisms behind the storage time improvement is certainly the redistribution and removal of heavy metals from the silicon. This is probably the dominant effect at lower temperatures (<1050 °C). At temperatures above 1100 °C the pronounced effect of the chlorine on the oxidation-induced stacking faults and on other crystalline defects deserves great attention and will be discussed in the next section.

Oxidation-induced stacking faults are extrinsic stacking faults formed during thermal oxidation of the silicon. They consist of an additional (111) plane bounded by Frank partial dislocations (Ravi and Varker 1974, Hu 1977). Their origin is associated with surface defects or grown-in bulk defects and they grow by emission of vacancies or by absorption of silicon interstitials. There is some evidence that the latter mechanism is more important in governing the OSFs behaviour (Hu 1977, Claeys *et al* 1978).

The growth and shrinkage of OSFs in a chlorine-containing ambient has been studied first for HCl by Shiraki (1975, 1976a, 1976b) and by Shibayama *et al* (1976). Hattori looked at the effect of TCE (Hattori 1976, 1977, 1978b) and Claeys *et al* studied C33 oxides (Claeys *et al* 1977, 1978). They all observed a strong influence on the OSF's growth kinetics.

Figure 16 shows the OSF length as a function of time with the C33 concentration as a parameter for oxidations at, respectively, 1100 °C, 1150 °C and 1200 °C (Claeys *et al* 1977). The stacking faults are made visible by means of a preferential etchant (Secco d'Aragona 1972, Wright Jenkins 1976). The length of the OSF refers to the intersection of the stacking fault with the surface as revealed by the selective etchant. From figure 16 it can be seen that at a constant temperature the initial growth rate of the OSFs is lower for C33 oxides than for standard dry oxides. The growth rate decreases with increasing C33 concentrations. The curves at 1150 °C show a growth and a shrinkage region. The

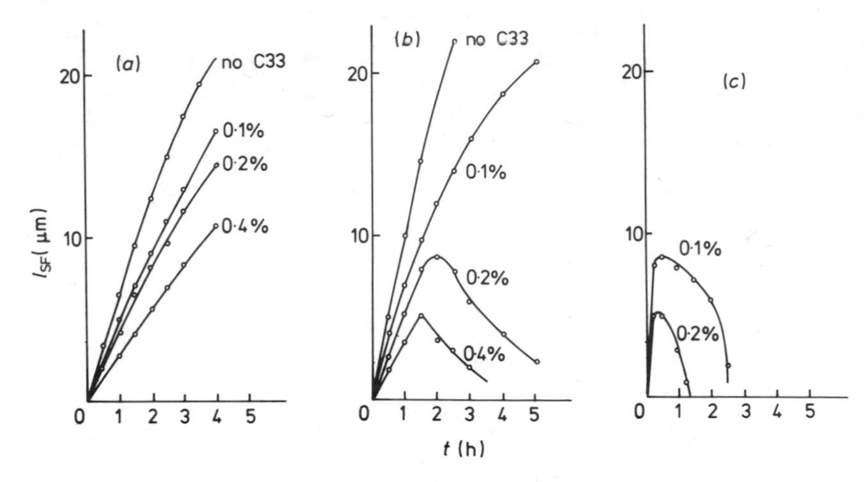

Figure 16. Stacking fault length as a function of time, with the C33 concentration as a parameter for oxidations at, respectively: (*a*) 1100 °C; (*b*) 1150 °C; and (*c*) 1200 °C (Claeys *et al* 1977).

maximum OSF length, and the oxidation time necessary to reach this maximum strongly depend on the C33 concentration. At 1200 °C the OSFs begin to shrink after a very short time, even for very small chlorine concentrations. For sufficiently long oxidation times, depending on the C33 concentration and oxidation temperature, it is possible to dissolve the OSFs completely.

The OSFs are believed to grow or shrink by absorption or emission of silicon interstitials. It is clear that this growth behaviour will be changed if an additional source or sink of interstitials is present in the vicinity of the defects. This is exactly what is going on during a chlorine oxidation. The total amount of silicon interstitials at the interface is strongly reduced by formation of Si—Cl bonds, leading to the formation of the second phase. At higher temperature, larger additions or longer oxidation times, more chlorine will be built up at the interface, creating a very efficient sink for silicon interstitials. This explains the OSF shrinkage phenomenon.

It has been reported by Shiraki (1976a) and by Hattori (1977, 1978b) that a chlorine oxidation not only reduces the length of existing OSFs but also removes the OSF nuclei. According to Shiraki a subsequent dry oxidation in a chlorine-free atmosphere does not generate OSFs, even if the first chlorine-containing oxide is removed before the reoxidation. Hattori, on the other hand, reports a much stronger OSF-inhibition effect if the initial chlorine-containing oxide is left intact (Hattori 1977).

Recently similar experiments have been carried out for reoxidations in dry or wet oxygen (Claeys *et al* 1979). The inhibition effect is illustrated in figure 17, indicating the occurrence of OSFs during a reoxidation in dry O_2, the duration of which is given in the abscissa. The ordinate gives the C33 concentration of the first chlorine oxidation. In figure 17 this oxide is left intact before the reoxidation. Claeys *et al* point out that such data should be regarded with great care. They show that prolonged preferential etching of the silicon for times up to 60 min does reveal the existence of OSFs lying deeper in the silicon bulk material. This is clearly demonstrated in figure 18 showing the OSF distribution after preferential Wright etching during respectively 3, 6, 12, 25, 45 or

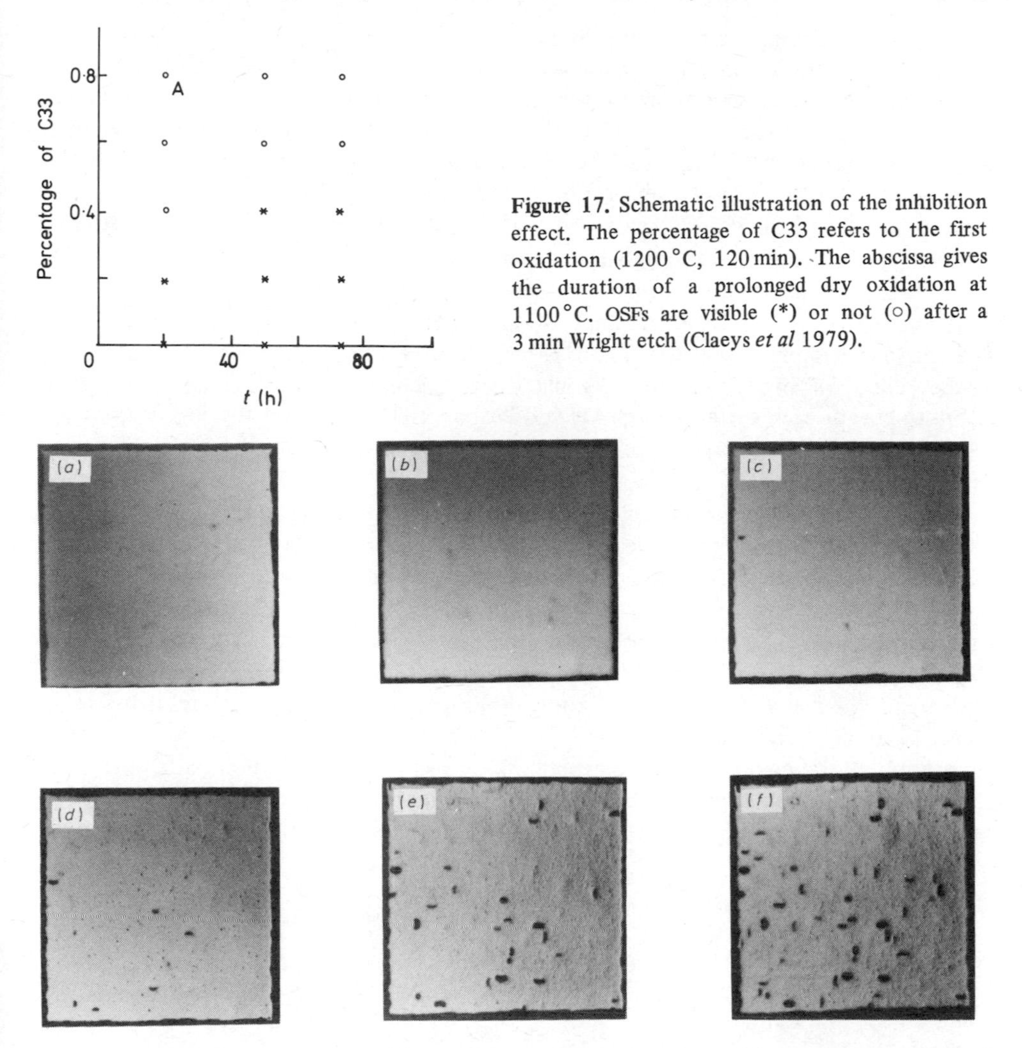

Figure 17. Schematic illustration of the inhibition effect. The percentage of C33 refers to the first oxidation (1200 °C, 120 min). The abscissa gives the duration of a prolonged dry oxidation at 1100 °C. OSFs are visible (*) or not (○) after a 3 min Wright etch (Claeys *et al* 1979).

Figure 18. The same area of a silicon wafer after a Wright etch of, respectively: (*a*) 3; (*b*) 6; (*c*) 12; (*d*) 25; (*e*) 45; and (*f*) 60 min. The oxidation conditions of this wafer are the same as for wafer A in figure 17 (Claeys *et al* 1979).

60 min. It should be emphasised that no OSFs were observed on this wafer (sample A from figure 17) after 3 min of etching. These authors also note that the inhibition effect is reduced by removal of the initial chlorine oxide or by subsequent oxidation in wet oxygen or steam.

The experiments described above indicate that the OSF inhibition effect is partly due to the presence of the chlorine itself, which has been built up in the oxide during the first oxidation. The chlorine concentration at the interface will decrease only very slowly during a second oxidation in dry oxygen, but will decrease much faster if the second oxidation is carried out in wet oxygen. On the other hand, the observation that part of the inhibition effect is still present if the chlorine oxide is removed before reoxidation, points to the assumption of an effective gettering of OSF nuclei. It is known that interstitial oxygen, present in the silicon bulk material, plays a very important role in the nucleation of OSFs, at least in Czochralski material, through the formation of SiO_2 precipitates. This brings us to an alternative explanation for the OSF inhibition effect as Pearce and Rozgonyi (1977) and Rozgonyi and Pearce (1977, 1978) prove that the interstitial oxygen concentration $[O_i]$ is reduced during a high-temperature HCl oxidation. The $[O_i]$, as determined from infrared absorption measurements, is decreased from 44 ppm to 25 ppm in 50 min at 1250 °C in O_2+0·5% HCl. At lower temperatures the gettering proceeds at a much slower rate. The lower $[O_i]$ reduces the probability of oxygen precipitation and inhibits the formation of OSFs.

Figure 19 shows a photomicrograph of an etched {111} cleavage cross section of a 500 μm thick sample annealed at 1250 °C for 4 h in O_2+0·5% HCl. The OSFs only occur in the central regions of the wafer, while the OSF-denuded zones extend from 100 to 125 μm beneath each surface. Rozgonyi and Pearce (1977) point out that the presence of OSFs deep in the bulk may be advantageous as they may act as gettering sinks during further processing.

It is clear now that the addition of chlorine to the high-temperature oxidation ambient strongly affects several properties of the silicon bulk material, such as density of oxidation-induced stacking faults, metallic precipitates and content of interstitial oxygen. A direct relation between enhanced leakage currents or dark current non-uniformities and the presence of OSFs has been reported by several groups (Shiraki *et al* 1971, Tanikawa *et al* 1976, Declerck *et al* 1976, Unter *et al* 1977, Janssens and Declerck 1978b, Ogden and Wilkinson 1977). It is also known that metallic impurities preferentially reside on the location of crystalline defects.

From the arguments above we would like to conclude that at high temperatures (>1100 °C) the elimination of OSFs is very important as being the driving force in the redistribution and/or subsequent removal of metallic impurities. As a consequence the

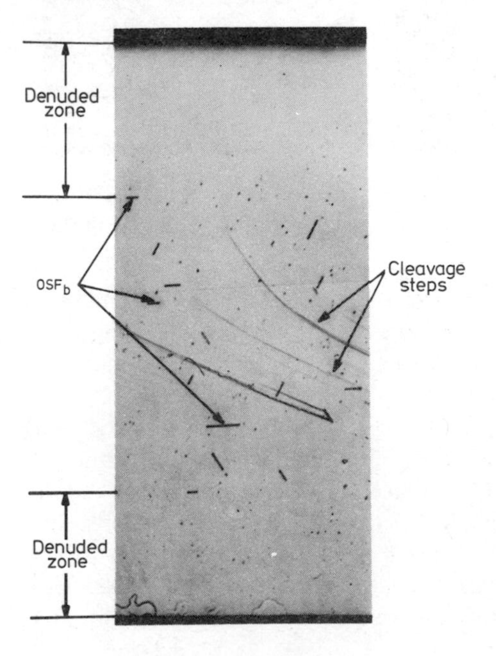

Figure 19. Photomicrograph of etched {111} cleavage cross section of a 500 μm thick sample annealed at 1250 °C for 4 h in O_2+0·5% HCl (Rozgonyi and Pearce 1977).

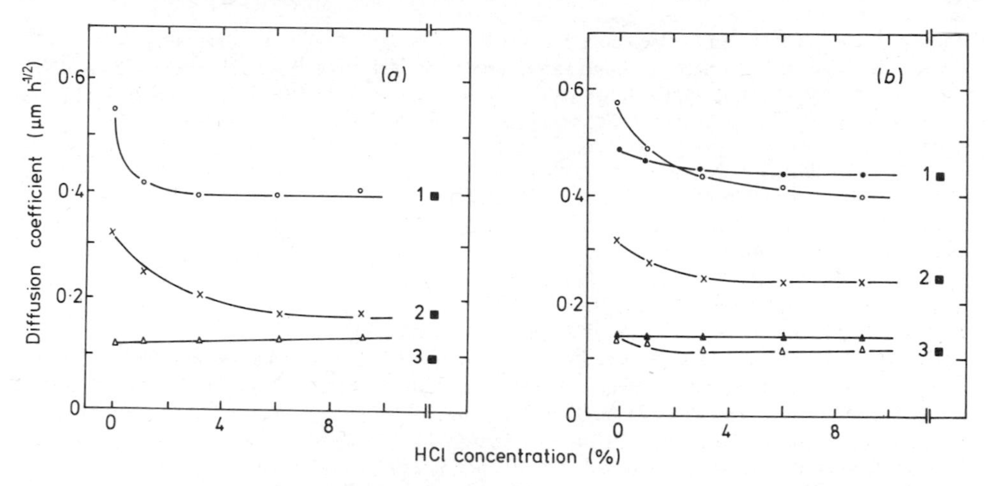

Figure 20. Diffusion coefficient of boron (*a*) and phosphorus (*b*): in silicon at different temperatures and HCl additions (Nabeta *et al* 1976); ○, at 1150°C (100); •, at 1150°C (111); ×, at 1100°C (100); △, at 1000°C (100); ▲, at 1000°C (111): and in N_2; 1■, at 1150°C; 2■, at 1100°C; 3■, at 1000°C.

storage time improvement observed during high-temperature chlorine oxidations should be attributed not only to a formation of metallic chlorides but also to the suppression of crystalline defects. At lower temperatures (<1050 °C) the effect on the OSFs is negligible and the removal of metallic impurities is the only mechanism left.

Finally it should be mentioned that the diffusion of dopants like phosphorus or boron is retarded in a chlorine-containing ambient as reported by Nabeta *et al* (1976). Their results are summarised in figure 20 and show that the oxidation-enhanced diffusion effect, which is normally observed for boron and phosphorus, completely disappears as soon as a few per cent HCl are added to the oxidation atmosphere. This observation again points to a reduction of the interstitial silicon concentration (or an enhancement of the vacancy concentration) at the surface, slowing down the OSF growth but also retarding the dopant diffusion.

7. Conclusion

The oxidation kinetics and the structure of chlorine oxides have been discussed. The physical and electrical properties of the oxide, of the interface, and of the silicon bulk material have been reviewed.

References

Baccarani G, Severi M and Soncini G 1973 *J. Electrochem. Soc.* **120** 1436–8

Balk P 1965 *Paper 111 presented at Electrochem. Soc. Meeting* Buffalo 1965

Baxter R D 1973 *Final Report contract* N00019-72C-0357 *by Battelle Memorial Institute, Columbus Laboratories, Ohio*

Butler S R, Feigl F J, Rohatgi A, Kraner H W and Jones K W 1977 *Paper 77 presented at Electrochem. Soc. Meeting* Philadelphia 1977

Chen M C and Hile J W 1972 *J. Electrochem. Soc.* **119** 223–5

Chou N J, Osburn C M, van der Meulen Y J and Hammer R 1973 *Appl. Phys. Lett.* **22** 380–1
Claeys C L, Laes E E, Declerck G J and van Overstraeten R J 1977 *Paper 224 presented at Electrochem. Soc. Meeting* Philadelphia 1977, also: *Semiconductor Silicon 1977* ed H R Huff and E Sirtl *The Electrochem. Soc. Softbound Symp. Ser.* 773–84
Claeys C L, Declerck G J and van Overstraeten R J 1978 *Paper 262 presented at Electrochem. Soc. Meeting* Seattle 1978, also: *Semiconductor Characterization Techniques* ed P A Barnes and G A Rozgonyi (Princeton: *The Electrochem. Soc.*) pp 366–75
—— 1979 *Paper 487 presented at Electrochem. Soc. Meeting* Los Angeles 1979
Deal B E 1978 *J. Electrochem. Soc.* **125** 576–9
Deal B E and Grove A S 1965 *J. Appl. Phys.* **36** 3770–8
Deal B E, Hess D W, Plummer J D and Ho C P 1978a *J. Electrochem. Soc.* **125** 340–6
Deal B E, Hurrle A and Schulz M J 1978b *J. Electrochem. Soc.* **125** 2024–7
Declerck G J, Hattori T, May G A, Beaudouin J and Meindl J D 1975 *J. Electrochem. Soc.* **122** 436–9
Declerck G J, De Meyer K M, Janssens E J, Laes E E, Van der Spiegel J and Claeys C L 1976 *Proc. Int. CCD Conf., Edinburgh 1976,* 23–30
Fogels E A and Salama C A T 1971 *J. Electrochem. Soc.* **118** 2002–6
Frenzel H, Singh B R, Haberle K and Balk P 1979 *Thin Solid Films* **58** 301–5
Goetzberger A, Klausmann E and Schulz M J 1976 *CRC Crit. Rev. Solid St. Sci.* **6** 1–43
Green J M, Osburn C M and Sedgwick T O 1974 *J. Electron. Mater.* **3** 579–99
Hattori T 1976 *J. Electrochem. Soc.* **123** 945–6
—— 1977 *Appl. Phys. Lett.* **30** 312–4
—— 1978a *Jap. J. Appl. Phys.* **17** 69–72
—— 1978b *Denki Kagaku* **46** 122–7
Heald D L, Das R M and Khosla R P 1976 *J. Electrochem. Soc.* **123** 302–3
Hirabayashi K and Iwamura J 1973 *J. Electrochem. Soc.* **120** 1595–601
Hu S M 1977 *J. Vac. Sci. Technol.* **14** 17–31
Janssens E J and Declerck G J 1976 *Solid State Devices (ESSDERC) 1975* (Paris: French Physical Society) pp 16–7
—— 1978a *J. Electrochem. Soc.* **125** 1696–703
—— 1978b *Paper 263 presented at Electrochem. Soc. Meeting* Seattle 1978, also: *Semiconductor Characterization Techniques* ed P A Barnes and G A Rozgonyi (Princeton: *The Electrochem. Soc.*) pp 376–85
Kooi E 1965 *Philips Res. Rep.* **20** 578–94
Kriegler R J 1972b *Appl. Phys. Lett.* **20** 449–51
—— 1972c *Thin Solid Films* **13** 11–4
—— 1973a *Semiconductor Silicon 1973, The Electrochem. Soc. Softbound Symp. Ser.* 363–75
Kriegler R J, Aitken A and Morris J D 1973b *Proc. 5th Conf. Solid State Devices, Tokyo* 1973; *Supplement to J. Jap. Soc. Appl. Phys.* **43** 1974 34–7
Kriegler R J, Cheng Y C and Colton D R 1972a *J. Electrochem. Soc.* **119** 388–92
Kriegler R J and Devenyi T F 1973c *Proc. 11th IEEE Symp. on Reliability Physics* pp 153–8
Kuhn M 1970 *Solid-St. Electron.* **13** 873–85
Linssen A J and Peek H L 1978 *Philips J. Res.* **33** 281–90
Mahajan S, Rozgonyi G A and Brasen D 1977 *Appl. Phys. Lett.* **30** 73–5
Mayo S and Evans W H 1977 *J. Electrochem. Soc.* **124** 780–5
Meek R L 1973 *J. Electrochem. Soc.* **120** 308–10
van der Meulen Y J and Cahill J G 1974 *J. Electron. Mater.* **3** 371–89
van der Meulen Y J, Osburn C M and Ziegler J F 1975 *J. Electrochem. Soc.* **122** 284–90
Monkowski J R 1978 *PhD Thesis* Pennsylvania State University
Monkowski J R, Tressler R E and Stach J 1978 *J. Electrochem. Soc.* **125** 1867–73
Montillo F and Balk P 1971 *J. Electrochem. Soc.* **118** 1463–8
Nabeta Y, Uno T, Kubo S and Tsukamoto H 1976 *J. Electrochem. Soc.* **123** 1416–7
Ogden R and Wilkinson J M 1977 *J. Appl. Phys.* **48** 412–4
Osburn C M 1974 *J. Electrochem. Soc.* **121** 809–15
Osburn C M and Ormond D W 1972a *J. Electrochem. Soc.* **119** 591–7
—— 1972b *J. Electrochem. Soc.* **119** 597–603

Pearce C W and Rozgonyi G A 1977 *Paper presented at Electrochem. Soc. Meeting* Philadelphia 1977, also: *Semiconductor Silicon 1977* ed H R Huff and E Sirtl *The Electrochem. Soc. Softbound Symp. Ser.* 606–15
Ravi K V and Varker C J 1974 *J. Appl. Phys.* **45** 263–71
Ritchey P M, Stach J and Tressler R E 1976 *Paper 324 presented at Electrochem. Soc. Meeting* Las Vegas 1976
Robinson P H and Heiman F P 1971 *J. Electrochem. Soc.* **118** 141–3
Rohatgi A, Butler S R, Feigl F J, Kraner H W and Jones K W 1977 *Appl. Phys. Lett.* **30** 104–6
Ronen R S and Robinson P H 1972 *J. Electrochem. Soc.* **119** 747–52
Rozgonyi G A, Mahajan S, Read M H and Brasen D 1976 *Appl. Phys. Lett.* **29** 531–3
Rozgonyi G A and Pearce C W 1977 *Appl. Phys. Lett.* **31** 343–5
—— 1978 *Appl. Phys. Lett.* **32** 747–9
Schroder D K and Guldberg J 1971 *Solid-St. Electron.* **14** 1285–97
Schroder D K and Nathanson H C 1970 *Solid-St. Electron.* **13** 577–82
Secco d'Aragona F 1972 *J. Electrochem. Soc.* **119** 948–51
Severi M and Soncini G 1972 *Electron. Lett.* **8** 402–4
Shibayama H, Masaki H, Ishikawa H and Hashimoto H 1976 *Appl. Phys. Lett.* **29** 136–8
—— 1976a *Jap. J. Appl. Phys.* **15** 1–10
—— 1976b *Jap. J. Appl. Phys.* **15** 83–6
Shiraki H 1976b *Jap. J. Appl. Phys.* **15** 83–6
Shiraki H, Matsui J, Kawamura T, Hanaoka M and Sasaki T 1971 *Jap. J. Appl. Phys.* **10** 213–7
Singh B R, Tyagi B D, Chandorkar A N and Marathe B R 1974 *Paper 42 presented at Electrochem. Soc. Meeting* San Francisco 1974
Singh B R and Balk P 1978 *J. Electrochem. Soc.* **125** 453–61
Stagg J P and Boudry M R 1978 *Revue Phys. Appl.* **13** 841–3
Tanikawa K, Ito Y and Sei H 1976 *Appl. Phys. Lett.* **28** 285–7
Tressler R E, Stach J and Metz D M 1977 *J. Electrochem. Soc.* **124** 607–9
Unter T F, Roberts P C T and Lamb D R 1977 *Electron. Lett.* **13** 93–4
Wright Jenkins M 1976 *Paper 118 presented at Electrochem. Soc. Meeting* Washington 1976
Young D R and Osburn C M 1973 *J. Electrochem. Soc.* **120** 1578–81
Zerbst M 1966 *Z. Angew. Phys.* **22** 30–3